S0-BXL-507

by

PAUL PENFIELD, JR.

Assistant Professor of Electrical Engineering
Massachusetts Institute of Technology

Frequency-Power Formulas

Published jointly by

The Technology Press of
The Massachusetts Institute of Technology

and

John Wiley & Sons, Inc., New York · London

TECHNOLOGY PRESS RESEARCH MONOGRAPHS

FREQUENCY-POWER FORMULAS
By Paul Penfield, Jr.

ELECTRONIC PROCESSES IN SOLIDS
By Pierre R. Aigrain

THE DYNAMIC BEHAVIOR OF THERMOELECTRIC DEVICES
By Paul E. Gray

HYDROMAGNETIC CHANNEL FLOWS
By Lawson P. Harris

PROCESSING NEUROELECTRIC DATA
By Walter A. Rosenblith and Members of the Communications Biophysics Group

MATHEMATICAL PROGRAMMING AND ELECTRICAL NETWORKS
By Jack B. Dennis

CIRCUIT THEORY OF LINEAR NOISY NETWORKS
By Hermann A. Haus and Richard B. Adler

NONLINEAR PROBLEMS IN RANDOM THEORY
By Norbert Wiener

Library of Congress Catalog Card Number: 60-16791

Printed in the United States of America

FOREWORD

There has long been a need in science and engineering for systematic publication of research studies larger in scope than a journal article but less ambitious than a finished book. Much valuable work of this kind is now published only in a semiprivate way, perhaps as a laboratory report, and so may not find its proper place in the literature of the field. The present contribution is the eighth of the Technology Press Research Monographs, which we hope will make selected timely and important research studies readily accessible to libraries and to the independent worker.

J. A. Stratton

Acknowledgment

This work was supported, in part, through the M.I.T. Electronic Systems Laboratory, by the Analysis and Design Branch, Power Division, Aeronautical Accessories Laboratory, Wright Air Development Division, Wright-Patterson Air Force Base, Ohio, under Contract AF 33(616)-3984, Project No. 8149, Task No. 61098, (Unclassified Title) "Advanced Analytical Studies." The work was also supported in part by a one-year Bendix Aviation Corporation fellowship, and, in part, through the M.I.T. Research Laboratory of Electronics, by the U.S. Army (Signal Corps), the U.S. Air Force (Office of Scientific Research, Air Research and Development Command), and the U.S. Navy (Office of Naval Research).

PREFACE

In 1956 J. M. Manley and H. E. Rowe of Bell Telephone Laboratories analyzed [60] power flow at various frequencies in a nonlinear capacitor. Their conclusions were the now-famous Manley-Rowe frequency-power formulas, which express a constraint on the powers entering the capacitor at each frequency. Their timing was fortunate, for within a year or so, the topic of parametric amplification was revived and applied to microwaves. Scarcely a paper on parametric amplifiers has appeared without mentioning the Manley-Rowe formulas, usually to show that the concept of parametric amplification is consistent with them.

It was clear from Manley and Rowe's work that the formulas also held for nonlinear inductors, but no mention was made of other devices of potential interest for frequency conversion. Although early parametric amplifiers [103, 114] were made from gyromagnetic material, it was not shown until 1960 that such material obeyed the general Manley-Rowe formulas. An interesting question is, 'What systems obey the Manley-Rowe formulas?' Here I shall report on an attempt to find necessary and sufficient conditions that a physical system obey the formulas.

My interest in frequency-power formulas started on April 23, 1958, when I attended an evening lecture (one in a series on solid state electronics, sponsored by the Boston Section of the I.R.E.) by Mr. A. G. Fox of Bell Telephone Laboratories, on masers and parametric amplifiers. The Manley-Rowe formulas were presented in simple terms to show the plausibility of parametric amplification.

It was clear from this talk that sets of coils, coupled together, obey the Manley-Rowe formulas. Reasoning that the formulas should hold for an induction motor, I decided to build a rotating-machine parametric amplifier [79]; later I was to learn that this device had been in use for many years under the name, "induction generator."

With an analogy shown between microwave parametric devices and rotating machines, it was natural to ask how far this analogy could be extended. How many devices can be used in this way? Specifically, what systems obey the Manley-Rowe frequency-power formulas?

A check of systems known to obey the formulas revealed that they were all conservative, in the classical mechanics sense. That is, they all had energy state functions. It was a simple matter to test the hypothesis that systems with energy functions obey the formulas: not only was this hypothesis correct, but the proof of the formulas was simplified by using, at the outset, the energy state function.

Meanwhile, Prof. Hermann A. Haus had also been interested in extending the Manley-Rowe formulas. His motivation was microwave amplification; he had already shown that the formulas held for a nonlinear electromagnetic medium, [35] and (for signals much smaller than the pump) a gyromagnetic medium [35] and a longitudinal electron beam [36]. Professor Haus agreed to supervise a study to determine necessary and sufficient conditions for the Manley-Rowe formulas to hold. (While Professor Haus was away from M.I.T. during the 1959-1960 academic year, Professor R. L. Kyhl took over as supervisor.)

The formulas of Manley and Rowe are not the only frequency-power formulas, however. Both Page [74] and Pantell [77] have given frequency-power formulas relating real power in nonlinear resistors. Manley and Rowe had shown formulas relating real power in nonlinear reactances, and reactive power in nonlinear resistors. It was not hard to find the fourth type of frequency-power formula, relating reactive power in reactances.

It was discovered that a modification of the formulas of each type allowed time-varying systems to be discussed. Distributed systems were more difficult, however, because the formulas should relate the power flow at the boundary of the system. This could be done for some systems, but apparently not for others. The key to this difficulty proved to be Hamilton's principle, a variational principle of classical mechanics. Distributed systems that obey Hamilton's principle also obey the Manley-Rowe formulas in such a way that the power flow can be evaluated (in a systematic way) at the boundary of the system. The study of a distributed system then reduces to the search for an appropriate Lagrangian. A Lagrangian for rotational fluid flow was obtained from Professor C.-C. Lin of the Mathematics Department in the Spring of 1959, and a suitable Lagrangian for gyromagnetic media was offered by Professor Alan L. McWhorter early in 1960.

The importance and simplicity of the Manley-Rowe formulas has led other workers to extend them. Many authors have discussed the formulas, interpreting them in new ways, proving them by new methods, and extending them. Of these, we mention S. Duinker [18] and P. A. Sturrock [99], each of whom independently discovered that systems described by Hamilton's canonical equations of motion obey the Manley-Rowe formulas.

I am pleased to acknowledge the great help given me by many people at M.I.T., including Professors D. C. White, R. L. Kyhl, P.A. Miles, L. J. Chu, H. H. Woodson, and H. A. Haus. The work reported here was done in the Energy Conversion Group at M.I.T. because of the liberal view of "energy conversion" held by those in charge, notably Professors White and Woodson.

This book is based on a thesis submitted in partial fulfillment of the requirements of the degree of Doctor of Science in the Department of Electrical Engineering at the Massachusetts Institute of Technology in June, 1960.

August 1960 Paul Penfield, Jr.

CONTENTS

PART ONE

THE FORMULAS

Chapter 1

INTRODUCTION

This book deals with frequency-power formulas, which constrain weighted sums of real or reactive powers entering a device at various frequencies, to be either zero, or positive, or negative. Although there are four different types of formulas, the most important are the Manley-Rowe formulas, which set a weighted sum of real powers at various frequencies entering a nonlinear, time-varying reactance, to zero.

A concept of fundamental importance is that of "power at each frequency." This concept is not difficult to grasp when applied to electrical devices, but it is more subtle for acoustic systems, electron beams, etc. Suppose we have an electrical terminal pair with voltage e(t) and current i(t) in the steady state. Here, as in the future, we mean by "steady-state operation" that the variables can be written as multiple Fourier series of the form

$$e(t) = e_0 + \mathrm{Re}\Sigma_\alpha e_\alpha e^{j\omega_\alpha t} \tag{1.1}$$

and

$$i(t) = i_0 + \mathrm{Re}\Sigma_\alpha i_\alpha e^{j\omega_\alpha t} \tag{1.2}$$

where α is an index. The subscripts α on the variables i_α, e_α indicate the peak values of the variables at frequency ω_α, given by formulas like

$$i_\alpha = 2\left\langle i(t)e^{-j\omega_\alpha t}\right\rangle \tag{1.3}$$

where the brackets denote a time average. The sums in Eqs. 1.1 and 1.2 include each frequency only once, not each frequency and its negative.

The time-average power through the terminal pair, P, is the average value of the product

$$e(t)i(t) \tag{1.4}$$

and hence can be written in the form

$$P = P_0 + \Sigma_\alpha P_\alpha \tag{1.5}$$

where

$$P_0 = e_0 i_0 \tag{1.6}$$

is called the "d-c power," and where

$$P_\alpha = \tfrac{1}{2}\mathrm{Re}\, e_\alpha i_\alpha^{*} \tag{1.7}$$

is called the "power at frequency ω_α," the asterisk indicating the complex conjugate. Notice that the instantaneous power $P(t) = e(t)i(t)$ can be written in a Fourier series like Eq. 1.1, but P_0, as defined in Eq. 1.6, is not the average value of $P(t)$, nor is P_α, as defined in Eq. 1.7, a Fourier coefficient of $P(t)$.

This decomposition of power into frequency "components" is quite common with electrical circuits, because the expression for instantaneous power, Eq. 1.4, is expressed as a well-defined product of two variables. Consider, however, the power-flow vector for an electron beam

$$\overline{P} = \overline{E} \times \overline{H} + \tfrac{1}{2} m \rho \overline{v}(\overline{v} \cdot \overline{v}) \tag{1.8}$$

where $\overline{E}$ is the electric field, $\overline{H}$ the magnetic field, m the electronic mass, ρ the electron density, and $\overline{v}$ the velocity. The second term of this formula, the mechanical part, does not appear as a simple product, so we are not sure how to break it apart. It could be broken into the two factors $\frac{\rho}{2}$ and $m\overline{v}(\overline{v} \cdot \overline{v})$, or other ways. It is not obvious which of these ways (if any) has physical significance.

Similar reasoning applies in defining the reactive power at each frequency, Q_α.

It is a property of linear time-invariant elements that no frequency conversion takes place in them. We are familiar with the classifications of such devices in terms of the real and reactive power at each frequency:

Lossless	$P_\alpha = 0$
Pure Lossy	$Q_\alpha = 0$
Passive	$P_\alpha \geq 0$
Inductive	$Q_\alpha \geq 0$
Capacitive	$Q_\alpha \leq 0$

(1.9)

These relations are conservation theorems for linear, time-invariant devices.

Such conservation theorems do not apply to nonlinear or time-varying devices. We cannot conclude from the fact that a nonlinear capacitor is lossless that all P_α vanish. Such a device is useful as a frequency converter only because we can put power in at some frequencies and get it out at other frequencies. The conservation theorems of Eq. 1.9 no longer hold, but might there not be milder conservation theorems that remain in the nonlinear case? There are: they are precisely the frequency-power formulas, of which the most famous are the Manley-Rowe equations.

The Manley-Rowe formulas [60] state that if the excitation of a nonlinear capacitor is such that the current and voltage have frequency components of the form $m\omega_1 + n\omega_0$, where m and n are integers, then

$$\sum_{m=-\infty}^{\infty} \sum_{n=1}^{\infty} \frac{nP_{mn}}{m\omega_1 + n\omega_0} = 0 \tag{1.10}$$

and

$$\sum_{m=1}^{\infty} \sum_{n=-\infty}^{\infty} \frac{mP_{mn}}{m\omega_1 + n\omega_0} = 0 \tag{1.11}$$

where P_{mn} is the power input at frequency $m\omega_1 + n\omega_0$. The result was originally proved for a nonlinear capacitor, but the extension to a nonlinear inductor was obvious. The nonlinear capacitor is lossless, so the sum of all P_{mn} must vanish. This is predicted by multiplying Eq. 1.10 by ω_0, multiplying Eq. 1.11 by ω_1, and adding. The Manley-Rowe formulas thus consist of the law of conservation of energy and one additional relation.

If the capacitor is terminated so that power flows only at frequencies ω_1, ω_0, and $\omega_- = \omega_1 - \omega_0$, the summations of Eqs. 1.10 and 1.11 reduce to

$$\frac{P_0}{\omega_0} = \frac{P_-}{\omega_-} = -\frac{P_1}{\omega_1} \tag{1.12}$$

and so the ratio of any two of the powers is the corresponding ratio of frequencies. Equations 1.12 are often used to show the plausibility of parametric amplification, in which power enters at frequency ω_1, and leaves at the other two frequencies.

These formulas also describe basically different devices. Weiss [115] has pointed out the similarity to the relations for a simple model of a three-level solid-state maser, which are

$$\frac{P_1}{\nu_1} = \frac{P_2}{\nu_2} = -\frac{P_3}{\nu_3} \tag{1.13}$$

where each P is the power input at the corresponding frequency ν, and where $\nu_1 + \nu_2 = \nu_3$. In addition, a differential gear described by the relation of the three shaft frequencies

$$\omega_1 = r(\omega_3 - \omega_2) \tag{1.14}$$

satisfies the torque relations [29]

$$r\tau_1 = \tau_2 = -\tau_3 \tag{1.15}$$

which can be expressed in terms of the power inputs at each shaft as

$$\frac{P_1}{\omega_3 - \omega_2} = \frac{P_2}{\omega_2} = -\frac{P_3}{\omega_3} \tag{1.16}$$

Furthermore, the theory of the balanced n-phase induction motor [116, Sec. 3.6.4; 25, Art. 9.2] predicts that the mechanical power output $-P_m$ is, in the absence of stator resistance, a fraction $(1-s)$ of the stator power input P_s. The remainder, sP_s, is dissipated in the rotor. But since s is the slip $\frac{\omega_r}{\omega_s}$, these results can be writt

$$\frac{P_m}{\omega_s - \omega_r} = \frac{P_r}{\omega_r} = -\frac{P_s}{\omega_s} \tag{1.17}$$

The similarity among Eqs. 1.12, 1.13, 1.16, and 1.17 is striking. It suggests that the Manley-Rowe formulas are still more general. We might ask how far the formulas can be extended — what devices obey them?

The motivation for asking this is strong. In the first place, people are interested in building practical frequency converters out of devices other than the simple nonlinear capacitor, for example, from ferrites [103, 114], electron beams [6, 56, 1, 55], and plasmas [47]. The Manley-Rowe formulas say something of importance about frequency conversion, so we naturally want to know what systems obey them.

Another motivation lies in the field of energy conversion. Although frequency-power formulas relate power at various frequencies, they are useful for energy converters in which one can identify the port through which power flows from its frequency. As noted before, balanced induction motors obey a limited form of the equations. Is it possible that unbalanced induction motors obey the more general form? What energy conversion devices can be used to circumvent the limitations on efficiency predicted by the Manley-Rowe formulas? What devices cannot be used?

Still another motivation lies in the application of the formulas to problems in hydrodynamic and magnetohydrodynamic stability. We cite two examples: Plasma confinement by RF fields appears quite promising [5, 113, 119], but as Haus [37] and Rostoker [88] have point out, there is a possibility of a parametric type of instability. And secondly, the excitation of cross waves by large-amplitude water waves [44] is an old [24] phenomenon undoubtedly due to parametric action.

Several other physical systems have displayed similar phenomena [71]. Parametric excitation has been observed in string vibrations [65, 82, 83] and pendulums [7, Sec. 45], and quite recently parametric energy conversion has been reported between electric and acoustic modes of yttrium iron garnet [98]. The interpretation of these phenomena would be clearer if the frequency-power formulas were known.

Several extensions of the Manley-Rowe formulas have been made. Duinker [19] has given an extension of a limited form of the formulas to a lumped, finite, reactive, electrical network. Rowe [89] has discussed an extension to a single time-varying reactance.

Haus has proved the formulas for some distributed systems of practical interest — the electromagnetic field in a nonlinear, dispersion-free, inhomogeneous, anisotropic, stationary, reciprocal medium [35]; the small-signal perturbation of a heavily-pumped gyromagnetic medium [35]; and the small-signal perturbation of a parametrically-pumped, longitudinal electron beam [36].

The broadest extension to date is that of Page [73], which shows that the formulas hold for any lossless system. Unfortunately this result is not true, as the counterexample of Appendix A shows. One must use a knowledge of the system beyond its losslessness to show the formulas to be valid. Some lossless devices that do not obey the formulas are the ideal rectifier, the switch, and the variable ideal transformer.

It is interesting that losslessness is not a sufficient condition for the formulas to hold. One would think that it would at least be necessary, but as the time-varying example of Sec. 4.4 shows, this is not true either.

We have attempted to find necessary and sufficient conditions for the Manley-Rowe formulas to hold for a system. To our knowledge, no non-trivial necessary conditions are known. However, any system that is described by an energy state function obeys the formulas. This sufficient condition is broad enough to cover cases of practical interest, as shown in Chapter 2.

The Manley-Rowe formulas are not the only type of frequency-power formulas, although they seem to be the most important. Manley and Rowe [60] have given another set of formulas, relating reactive power at various frequencies in a nonlinear resistor. These are

$$\Sigma_m \Sigma_n m Q_{mn} = 0 \tag{1.18}$$

and

$$\Sigma_m \Sigma_n n Q_{mn} = 0 \tag{1.19}$$

where each Q_{mn} is the reactive power into the resistor at frequency $m\omega_I + n\omega_0$. These formulas have not been very useful in the past.

More useful formulas are those that relate real power at various frequencies in nonlinear resistors. Such formulas have been given by Page [74, 75] and Pantell [77]. They are inequalities, rather than equalities; for example, Pantell's formulas are

$$h_m = \Sigma_m \Sigma_n m^2 P_{mn} \geq 0 \quad (1.20)$$

and

$$h_n = \Sigma_m \Sigma_n n^2 P_{mn} \geq 0 \quad (1.21)$$

They hold for nonlinear resistors for which the incremental resistance

$$R_i = \frac{de}{di} \quad (1.22)$$

is nonnegative. We call such inequalities formulas of Type III, whereas the Manley-Rowe formulas are Type I, and the formulas relating reactive power in nonlinear resistors are Type II.

Thus far we have seen formulas corresponding to three of the four boxes in the chart below

	Real Power	Reactive Power
nonlinear reactance	Type I Manley and Rowe Equalities	
nonlinear resistance	Type III Page Pantell Inequalities	Type II Manley and Rowe Equalities

The remaining formulas, of Type IV, relate reactive power in nonlinear reactances; they are derived and discussed in Chapter 3.

In Chapter 2 we show that the energy-function method of proving the Manley-Rowe formulas is the logical generalization of both the previously known methods. In Chapter 3 the frequency-power formulas of the four types are proved for devices that are not only nonlinear, but also time-varying. One such time-varying device of importance is the switch, which obeys formulas of Types II and III.

The succeeding three chapters constitute Part II of the book. They are concerned with specific systems. Lumped systems that obey the first and/or fourth types of formulas are discussed in Chapter 4, and lumped systems that obey the second and third types are discussed in Chapter 5. Distributed systems present a special challenge, because we want to evaluate the expressions for power on the boundary of the system at the ports. In Chapter 6 it is shown that this is possible for many systems of practical interest, including all distributed systems that obey Hamilton's principle.

Some applications of the frequency-power formulas are given in Part III. The formulas are conservation principles, and are of the most value when used as such — as an aid to thinking and understanding, as a check on analytic results, as a tool for finding fundamental limits, etc. Nevertheless, there are problems in which the frequency-power formulas can save many steps, and sometimes even lead directly to the desired results. Generally speaking, this happens when the number of frequencies present is small. Some applications of this sort to rotating machines are discussed in Chapter 7, and some communications applications in Chapter 8. We cannot pretend to discuss these applications exhaustively.

Frequency-power formulas of each type, under a variety of frequency constraints, are collected for reference in Appendix B, along with a partial list of physical devices known to obey them. In Appendix C it is shown that for the purpose of applying formulas of Type I (Manley-Rowe formulas), one may lump together power at all harmonics of a given frequency, considering it to be power at the fundamental (or at any one of the harmonics). In Appendix D it is shown that for a rotating shaft, one can consider the d-c power to be power at a frequency equal to the average shaft speed. And in Appendix E it is shown under what conditions new information is obtained by using a different choice of independent frequencies.

To help guide the reader, we can say that the most important points to be made are:

(1) The Manley-Rowe formulas hold for any physical system that has an energy-state function.

(2) Distributed systems that obey Hamilton's principle also obey the Manley-Rowe formulas.

(3) There are four types of frequency-power formulas, and they ought to have equal theoretical standing, in spite of the fact that, of the four, the Manley-Rowe formulas (Type I) have the most practical importance.

(4) Frequency-power formulas of each type apply (in a restricted form) to time-varying devices, as well as to time-invariant nonlinear devices.

(5) Frequency-power formulas are conservation principles, and are of the most value when used as such.

Chapter 2

THE ENERGY-FUNCTION METHOD

Only three essentially different ways of proving the Manley-Rowe formulas, Eqs. 1.10 and 1.11, have been published. One of these is applicable only to linear time-varying reactive systems; so there are two methods for a nonlinear system. Both of these methods can be generalized to multiport nonlinear systems, but the natural extension of each uses conditions that are sufficient to enable one to define an energy state function for the system. In Chapter 3 we prove the formulas by using the energy state function at the outset; this method can be considered as a logical generalization of both the previously known methods.

The state function method is preferable because it is general, yet simple; because it provides a relatively simple criterion (the existence of an energy state function); and because time-varying systems are treated very easily.

The two previous methods of proof of the Manley-Rowe formulas are discussed in the first two sections of this chapter, and it is shown in the third section that the logical extensions of these methods lead to an energy state function for the system. The actual use of this energy function is shown in Chapter 3.

2.1 The Manley-Rowe Formulas

The Manley-Rowe formulas were originally proved for a nonlinear time-invariant capacitor, but by simple analogy it was known the results held for other single energy-storage elements. It is assumed (for the full equations) that there is no loss or hysteresis, and that the voltage e is some function of the charge q

$$e = e(q) \tag{2.1}$$

It is further assumed that the excitation is such that power flows into or out of the capacitor only at frequencies of the form $m\omega_1 + n\omega_0$. We call P_{mn} the power that flows <u>into</u> the capacitor at frequency $m\omega_1 + n\omega_0$.

Because of the nonlinearity, there can be net power into the capacitor at some frequencies, and out at other frequencies, in spite of the fact that the total power input $\Sigma_m \Sigma_n P_{mn}$ vanishes. There is, however, one constraint caused by the fact that ω_1 and ω_0 can be specified independently. The easiest way to state this constraint is to write it together with the law of conservation of energy, in the for

$$\Sigma_m \Sigma_n \frac{mP_{mn}}{m\omega_1 + n\omega_0} = 0 \qquad (2.2)$$

and

$$\Sigma_m \Sigma_n \frac{nP_{mn}}{m\omega_1 + n\omega_0} = 0 \qquad (2.3)$$

These are the Manley-Rowe formulas.

To prove these formulas by any method, we write the voltage in the form of a multiply-periodic Fourier series

$$e(t) = e_0 + \text{Re}\, \Sigma_m \Sigma_n E_{mn} e^{j(m\omega_1 + n\omega_0)t} \qquad (2.4)$$

where ω_1 and ω_0 are independent frequencies. To say that two frequencies are independent, however, is equivalent to saying that their two phases are independent, so let us re-write Eq. 2.4 in the form

$$e(t) = e_0 + \text{Re}\, \Sigma_m \Sigma_n E_{mn} e^{j(m\phi_1 + n\phi_0)} \qquad (2.5)$$

where

$$\phi_1 = \omega_1 t \qquad (2.6)$$

and

$$\phi_0 = \omega_0 t \qquad (2.7)$$

are independent phases. We also represent in this form the charge q and the current i :

$$q(t) = q_0 + \text{Re}\, \Sigma_m \Sigma_n Q_{mn} e^{j(m\phi_1 + n\phi_0)} \qquad (2.8)$$

and

$$i(t) = \text{Re}\, \Sigma_m \Sigma_n I_{mn} e^{j(m\phi_1 + n\phi_0)} \qquad (2.9)$$

where the current and charge Fourier coefficients I_{mn} and Q_{mn} are related by

$$I_{mn} = j(m\omega_1 + n\omega_0)Q_{mn} \qquad (2.10)$$

because the current is the time-derivative of the charge. In Eqs. 2.2, 2.3, 2.4, 2.5, 2.8, and 2.9, the summations over m and n are intended to include each frequency, but not its negative. Thus if the integers (m, n) are included in the sum, the set of integers $(-m, -n)$ is not. Furthermore, the sum is not intended to include the d-c term, which is shown separately, so $(m = 0, n = 0)$ is not included. A simple way to insure that the summation is taken properly is to sum only over positive frequencies, or over one of the indices (m or n) from 0 to ∞, and the other from -∞ to ∞, omitting (m = 0, n = 0).

The Fourier coefficients like E_{mn} are the peak values of the components of the variables at frequency $m\omega_1 + n\omega_0$. They are evaluated by averages of the form

$$E_{mn}{}^* = 2\left\langle e(\phi_1, \phi_0)\, e^{j(m\phi_1 + n\phi_0)}\right\rangle \tag{2.11}$$

where the asterisk indicates the complex conjugate, and where the brackets indicate an average with respect to ϕ_1 and ϕ_0. Note that we consider the voltage and the charge as functions of the phases, through Eqs. 2.5 and 2.8.

The Manley-Rowe formulas can now be proven easily, by the method originally used [60] (rather, a slight modification of it). We calculate from Eq. 2.8

$$\frac{\partial q(\phi_1, \phi_0)}{\partial \phi_1} = \mathrm{Re}\, \Sigma_m \Sigma_n\, jmQ_{mn} e^{j(m\phi_1 + n\phi_0)} \tag{2.12}$$

and so

$$\left\langle e\, \frac{\partial q}{\partial \phi_1}\right\rangle = \tfrac{1}{2}\mathrm{Re}\, \Sigma_m \Sigma_n\, jmQ_{mn} E_{mn}{}^* \tag{2.13}$$

by Eq. 2.11. The left-hand side of Eq. 2.13 vanishes when averaged with respect to ϕ_1, for it becomes the integral of $e\,dq$, which is a perfect differential, evaluated at ϕ_1 and $\phi_1 + 2\pi$, or for identical values of q. Thus the right-hand side can be equated to zero in the form

$$\Sigma_m \Sigma_n \frac{mP_{mn}}{m\omega_1 + n\omega_0} = 0 \tag{2.14}$$

where

$$P_{mn} = \tfrac{1}{2}\mathrm{Re}\, E_{mn}{}^* I_{mn} \tag{2.15}$$

is the power input to the capacitor at frequency $m\omega_1 + n\omega_0$. This is one of the Manley-Rowe formulas; the other is obtained in a similar way.

We have shown here the method originally used by Manley and Rowe [60] to prove their formulas. In Sec. 2.2 we discuss the other important method of proof, and in Sec. 2.3 we show that the energy-function method of Chapter 3 can be considered the logical generalization of both of them.

2.2 Alternate Methods

Essentially only two alternate methods of proving the Manley-Rowe formulas have been given. Weiss [115] has shown that the equations also hold for a simplified model of a three-level, solid-state maser, but his argument involves the quantum nature of the maser process, and is not a method of proving the Manley-Rowe formulas.

One of the alternate methods was used by Duinker [19] in his study of linear time-varying reactances for frequency conversion. He did not obtain both Manley-Rowe formulas, but only one. By calculating the actual impedance matrix of a reactive time-varying network, Duinker found that it satisfied one formula of the form of Eq. 2.2. He postulated (correctly) that the full form of the Manley-Rowe formulas, both Eqs. 2.2 and 2.3, should hold for a network of nonlinear reactive elements, but was not able to show this by his method. (This fact is shown in Sec. 4.10, and was also derived by Duinker himself [18]. Since this method is not suitable for nonlinear devices, we will not discuss it further.

The other method of proof has been used by Page [73], Salzberg [90], Carroll [12], Clavier [16], and Callen [10]. We write the law of conservation of energy

$$\Sigma_m \Sigma_n P_{mn} = 0 \tag{2.16}$$

in the form

$$\omega_1 \Sigma_m \Sigma_n \frac{mP_{mn}}{m\omega_1 + n\omega_0} + \omega_0 \Sigma_m \Sigma_n \frac{nP_{mn}}{m\omega_1 + n\omega_0} = 0 \tag{2.17}$$

where we have used only the fact that the nonlinear capacitor is lossless. It is clear that the Manley-Rowe formulas are valid provided we set each of the two sums individually to zero.

Page [73] reasoned that since ω_1 and ω_0 are independent frequencies, they can be varied independently, and therefore the factor that multiplies each in Eq. 2.17 must vanish. Thus all lossless devices should obey the Manley-Rowe formulas. This, however, overlooks the possibility that the sums might themselves depend on ω_1 and ω_0: when one of these is varied, the sums might vary also. Some lossless devices that do not obey the Manley-Rowe formulas are the ideal diode, the switch, the variable ideal transformer, and their mechanical counterparts, including the ideal ratchet and the periodically-applied clamp.

Salzberg [90], on the other hand, noted that the ratio of P_{mn} to $m\omega_1 + n\omega_0$ is, from Eq. 2.15,

$$\frac{P_{mn}}{m\omega_1 + n\omega_0} = \tfrac{1}{2} \operatorname{Re} jE_{mn}{}^* Q_{mn} \tag{2.18}$$

He then stated that the Q_{mn} are functions of the E_{mn} and the nonlinear characteristic of the capacitor, but not of ω_1 or ω_0. Although the frequency of which Q_{mn} is the Fourier component does depend on ω_1 and ω_0, the value of Q_{mn} does not. Salzberg then was able to conclude that the only admissible solution of Eq. 2.17 for various values of ω_1 and ω_0 is with each sum individually vanishing. Although the expressions in his paper [90] are for quite simple cases, with only a few frequencies, his method (as he pointed out) is also used to obtain the full formulas.

This method has been given more recently, but without significant improvement. Carroll [12] used essentially Salzberg's reasoning, but in more mathematical form. Clavier [16] derived Eq. 2.17 and then claimed that this implied Eqs. 2.2 and 2.3, but unfortunately did not disclose his reasoning in detail. Callen [10] also derived Eq. 2.17 and demonstrated with a simple example that Salzberg's reasoning is correct — the Q_{mn} and E_{mn} are related by the capacitor, independent of ω_1 and ω_0.

Manley and Rowe [59] have commented on this method, pointing out (correctly) that going from Eq. 2.17 to Eqs. 2.2 and 2.3 does require knowledge of the device beyond its losslessness. They consider an ensemble of systems with different ω_1 and ω_0, but the same E_{mn} and Q_{mn}.

As Manley and Rowe [59] have stated, this alternate method does not have any analytical advantage over the original method. The contribution, then, is one of providing insight, not rigor.

2.3 Extension of These Methods

Suppose we wish to extend the method of Manley and Rowe, as given in Sec. 2.1, to multiport devices. To demonstrate the logical extension, let us consider two coils in a nonlinear medium, with coupling. We suppose there is no winding resistance, hysteresis, eddy currents, or other loss. The constitutive relations of the system then are in the form

$$i_1 = i_1(\lambda_1, \lambda_2) \tag{2.19}$$

and

$$i_2 = i_2(\lambda_1, \lambda_2) \tag{2.20}$$

where the subscripts 1 and 2 refer to the coils, and where the λ variables are the flux linkages, the time-integrals of the voltages. We write each of the variables in Fourier series

$$\lambda_1(t) = \lambda_{10} + \operatorname{Re} \Sigma_m \Sigma_n \Lambda_{1mn} e^{j(m\phi_1 + n\phi_0)} \tag{2.21}$$

$$\lambda_2(t) = \lambda_{20} + \operatorname{Re} \Sigma_m \Sigma_n \Lambda_{2mn} e^{j(m\phi_1 + n\phi_0)} \tag{2.22}$$

$$i_1(t) = i_{10} + \operatorname{Re} \Sigma_m \Sigma_n I_{1mn} e^{j(m\phi_1 + n\phi_0)} \tag{2.23}$$

$$i_2(t) = i_{20} + \operatorname{Re} \Sigma_m \Sigma_n I_{2mn} e^{j(m\phi_1 + n\phi_0)} \tag{2.24}$$

$$e_1(t) = \text{Re}\, \Sigma_m \Sigma_n E_{1mn} e^{j(m\phi_1 + n\phi_0)} \tag{2.25}$$

and

$$e_2(t) = \text{Re}\, \Sigma_m \Sigma_n E_{2mn} e^{j(m\phi_1 + n\phi_0)} \tag{2.26}$$

where we show each variable as a function of the phases ϕ_1 and ϕ_0. Since the voltages are the time-derivatives of the charges, each E_{1mn} and E_{2mn} is $j(m\omega_1 + n\omega_0)$ times the corresponding Λ_{1mn} or Λ_{2mn}.

Just as in Sec. 2.1 we form the partial derivatives

$$\frac{\partial\lambda_1(\phi_1, \phi_0)}{\partial\phi_1} = \text{Re}\, \Sigma_m \Sigma_n\, jm\Lambda_{1mn} e^{j(m\phi_1 + n\phi_0)} \tag{2.27}$$

and

$$\frac{\partial\lambda_2(\phi_1, \phi_0)}{\partial\phi_1} = \text{Re}\, \Sigma_m \Sigma_n\, jm\Lambda_{2mn} e^{j(m\phi_1 + n\phi_0)} \tag{2.28}$$

and take the averages

$$\left\langle i_1 \frac{\partial\lambda_1}{\partial\phi_1} \right\rangle = \tfrac{1}{2}\text{Re}\, \Sigma_m \Sigma_n\, jm\Lambda_{1mn} I_{1mn}^* \tag{2.29}$$

and

$$\left\langle i_2 \frac{\partial\lambda_2}{\partial\phi_1} \right\rangle = \tfrac{1}{2}\text{Re}\, \Sigma_m \Sigma_n\, jm\Lambda_{2mn} I_{2mn}^* \tag{2.30}$$

using expressions for I_{1mn}^* and I_{2mn}^* similar to Eq. 2.11. We may then add Eqs. 2.29 and 2.30, obtaining

$$\left\langle i_1 \frac{\partial\lambda_1}{\partial\phi_1} + i_2 \frac{\partial\lambda_2}{\partial\phi_1} \right\rangle = \Sigma_m \Sigma_n \frac{mP_{mn}}{m\omega_1 + n\omega_0} \tag{2.31}$$

where

$$P_{mn} = \tfrac{1}{2}\text{Re}\, (I_{1mn}^* E_{1mn} + I_{2mn}^* E_{2mn}) \tag{2.32}$$

is the power input to the device at frequency $m\omega_1 + n\omega_0$.

To show that the Manley-Rowe formulas hold, we want to show that the left-hand side of Eq. 2.31 vanishes. At this point in Sec. 2.1, we noted that the average with respect to ϕ_1 of the corresponding quantity vanished, since the quantity $e\,dq$ was a perfect differential. Here we need to show that $i_1 d\lambda_1 + i_2 d\lambda_2$ is a perfect differential. This is true if and only if

$$\frac{\partial i_1(\lambda_1, \lambda_2)}{\partial \lambda_2} = \frac{\partial i_2(\lambda_1, \lambda_2)}{\partial \lambda_1} \quad (2.33)$$

The logical extension of the proof of Sec. 2.1 then requires that Eq. 2.33 hold.

But Eq. 2.33 is a condition that is sufficient [42, Section 6.10; 80] to enable us to define an energy state function (the stored magnetic energy) by the line integral in the $\lambda_1 - \lambda_2$ plane

$$W_m(\lambda_1, \lambda_2) = \int i_1 d\lambda_1 + \int i_2 d\lambda_2 \quad (2.34)$$

The proof of the Manley-Rowe formulas is simplified by working with the energy state function from the beginning. This "energy function method" is given in Chapter 3; it is seen to be a logical extension of the original proof of Manley and Rowe [60].

Now suppose that we wish to extend the alternate proof given in Sec. 2.2 to multiport devices. We shall see that this involves assumptions that are also sufficient to enable us to define an energy state function.

We discuss the same device — the nonlinearly-coupled coils, with the currents i_1 and i_2 functions of the flux linkages λ_1 and λ_2. Then, because the system is lossless, we write

$$\omega_1 \Sigma_m \Sigma_n \frac{mP_{mn}}{m\omega_1 + n\omega_0} + \omega_0 \Sigma_m \Sigma_n \frac{nP_{mn}}{m\omega_1 + n\omega_0} = 0 \quad (2.35)$$

where

$$\frac{P_{mn}}{m\omega_1 + n\omega_0} = \mathrm{Re}\, j(I_{1mn}{}^*\Lambda_{1mn} + I_{2mn}{}^*\Lambda_{2mn}) \quad (2.36)$$

Suppose the system is operating with given Λ_{1mn}, Λ_{2mn}, λ_{10}, λ_{20}, ω_1, and ω_0. Then in particular, $\lambda_1(t)$ and $\lambda_2(t)$ are determined, and hence so are $i_1(t)$ and $i_2(t)$, and hence so are I_{1mn} and I_{2mn}. Now suppose that ω_1 is changed slightly, keeping λ_{10}, λ_{20}, Λ_{1mn}, and Λ_{2mn} the same. Does this change affect the I_{1mn} or I_{2mn} at all? It does not. By definition I_{1mn} or I_{2mn} are independent of time. But in the specification of $\lambda_1(t)$ and $\lambda_2(t)$, and therefore in the specification of $i_1(t)$ and $i_2(t)$, ω_1 appears only multiplied by the time t. In any functional dependence ω_1 can only appear multiplied by t, and therefore any dependence of I_{1mn} or I_{2mn} must contain t if it is to contain ω_1. Thus the I_{1mn} and I_{2mn} cannot depend on ω_1, but only on the Fourier components of the flux linkages. It is true that the frequency of which I_{1mn} is the Fourier coefficient changes, but the value of I_{1mn} does not.

It is then clear from Eq. 2.36 that the sums of Eq. 2.35 are independent of ω_1. If we change ω_1 slightly, the left-hand side can vanish only if the first summation itself vanishes: Thus we have shown that each sum individually vanishes, or the Manley-Rowe formulas hold.

Notice that this proof, an extension of the method of Salzberg and others [90, 12, 16, 10, 59] requires three assumptions:

(1) The currents depend at each instant on the flux linkages, as expressed in Eqs. 2.19 and 2.20.
(2) The device is lossless.
(3) The variables $\lambda_1(t)$ and $\lambda_2(t)$ are physically independent, so that each can be specified as a function of time independent of the other.

This last condition is necessary if we are to vary ω_1 without thereby affecting the flux linkage Fourier coefficients.

But these three conditions are sufficient to define an energy state function of the form of Eq. 2.34, for as we shall now show, they imply Eq. 2.33.

Let us operate the system around the path in the $\lambda_1 - \lambda_2$ plane shown in Fig. 2.1. This operation is possible since the variables

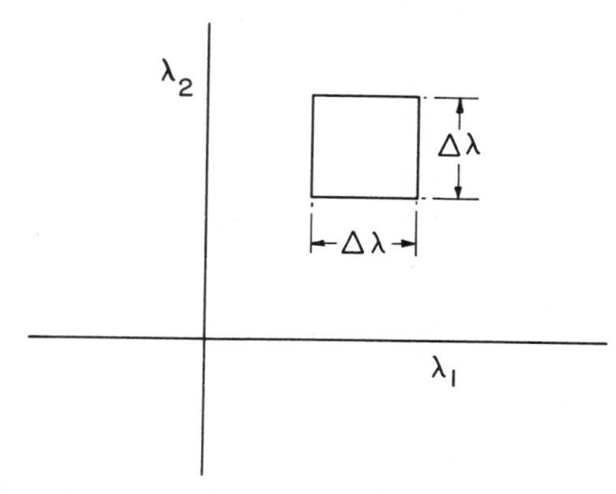

Fig. 2.1. Path of operation in the $\lambda_1 - \lambda_2$ plane. The position of the path is arbitrary, so the result of Eq. 2.33 holds for any values of λ_1 and λ_2

λ_1 and λ_2 are independent. Since the system is lossless, the net energy out over such a cycle must vanish, for otherwise we could get time-average power out of the device by merely repeating this cycle periodically. The net energy out of the system per cycle is the line integral around the path in the $\lambda_1 - \lambda_2$ plane

$$-\int i_1 d\lambda_1 - \int i_2 d\lambda_2 \tag{2.37}$$

which is, to second order in the quantity $\Delta\lambda$, merely

$$(\Delta\lambda)^2 \left[\frac{\partial i_2}{\partial \lambda_1} - \frac{\partial i_1}{\partial \lambda_2} \right] \tag{2.38}$$

shown either by direct calculation or by use of Stoke's Law. Because this vanishes at each point in the $\lambda_1 - \lambda_2$ plane, Eq. 2.33 must hold. Therefore we may define an energy state function of the form of Eq. 2.34.

Both methods [60, 90] of proving the Manley-Rowe formulas, when extended to multiport devices, lead logically to the existence of an energy state function. Thus the energy function method we use in Chapter 3 can be considered a logical extension of each of the methods of proof.

The energy function method leads to general results with mathematical simplicity, and provides some insight into what sort of systems obey the Manley-Rowe formulas. The criterion of having an energy state function is one that is easily applied to specific systems. Time-varying systems are merely those that have an energy state function with an explicit time dependence; they are discussed easily by the energy function method.

Chapter 3

FOUR TYPES OF FREQUENCY-POWER FORMULAS

In this chapter we shall prove the four types of frequency-power formulas. The first type (Manley-Rowe formulas) will be proved by an energy function method; the formulas are extended in a variety of ways:

(1) We allow more general frequencies than those of the form $m\omega_1 + n\omega_0$.
(2) We allow for more than two independent frequencies.
(3) We take into account d-c power input (by a modification of the formulas), such as that at the shaft of a rotating machine.
(4) We prove the formulas starting from a simple, general characteristic (the existence of an energy state function), and so cover many important physical systems.
(5) We include time-varying systems.

Of these extensions, some are trivial. It is easy to account for more general frequency relations and more than two independent frequencies [18, 99, 12, 120]; it is only for reasons of simplicity that not all previous writers have done so. We use notation so general that it will be no bother for us to make this extension.

Other extensions are of some importance. The extension to include d-c power input has been useful with rotating machines, in which d-c mechanical power flows. The extension to other physical systems, which we will discuss in Chapters 4 and 6, is also important.

In Sec. 3.1 we lead up to the proofs of the four types of frequency-power formulas, by introducing the notation used. The actual proofs are contained in Secs. 3.2, 3.3, 3.4, and 3.5. Appendix B includes a summary of the four types of formulas, together with their forms for some special cases, and a tabulation of some of the devices known to obey them.

3.1 Notation

Generally speaking, frequency-power formulas of Types I and IV hold for nonlinear time-varying reactances, and frequency-power formulas of Types II and III hold for nonlinear time-varying resistances. Some devices obey formulas of only one type, and some obey all four types, so the preceding statement must not be taken too literally. But in general, reactive, energy-storage devices obey Types I and IV, and resistive, dissipative devices obey Types II and III. The choice of notation in each of the four proofs to follow is clear if this is kept in mind.

Two important notational schemes are used in this chapter. The first is for the variables that describe the devices, either reactive or dissipative; the second is for the frequencies.

The Variables. Although the most important applications of frequency-power formulas are to electrical devices, we keep the notation for variables general. We suppose there are two variables that, when multiplied together, give power entering a device, or vector power flow density. Typical variables are voltage e and current i; mechanical force f and velocity v; rotational torque τ and speed ω; electrochemical potential μ and particle current density $\bar{\iota}$; electric and magnetic fields forming Poynting's vector $\overline{E} \times \overline{H}$; and acoustic pressure π and velocity $\overline{v}$. The general notation for such variables will be f_i and v_i (taken from the mechanical case), where i is an index.

Dissipative devices are characterized by relations among these variables: a nonlinear resistor has a voltage-current curve; a mechanical friction device has a nonlinear force-velocity curve, and a dissipative medium would have a current density $\overline{J}$ (related to $\overline{H}$ through one of Maxwell's equations) a function of $\overline{E}$. Thus we say for dissipative systems that

$$f_i = f_i(v_k) \tag{3.1}$$

or that each f_i is a function of all the v_k.

If these "constitutive" relations, Eqs. 3.1, are "reciprocal," in the sense that

$$\frac{\partial f_i(v_k)}{\partial v_\ell} = \frac{\partial f_\ell(v_k)}{\partial v_i} \tag{3.2}$$

then we may define a "dissipation function" by the line integral in multidimensional v_i space [42, Sec. 6.10; 80]

$$G(v_k) = \Sigma_i \int f_i dv_i \tag{3.3}$$

where

$$f_i = \frac{\partial G}{\partial v_i} \tag{3.4}$$

In other contexts this function has been called the Rayleigh dissipation function [30, Sec. 1.5], the content [68], and the co-content [68]. If Eq. 3.2 does not hold, we cannot define this state function, for the integral of Eq. 3.3 depends on the path chosen.

Time-varying dissipative systems are those for which the constitutive relations of Eq. 3.1 are replaced by ones that depend explicitly on time:

$$f_i = f_i(v_k, t) \qquad (3.5)$$

If Eq. 3.2 holds for the time-varying dissipative system, then the state function defined in Eq. 3.3 will be time-varying, that is, it will have an explicit time dependence.

A typical device that has a time-varying constitutive relation is a variable resistor (rheostat) whose shaft is being moved as a function of time, so that the resistance is some function $R(t)$ of time. Then the constitutive equation is

$$e(i, t) = R(t)i \qquad (3.6)$$

and the voltage is seen to be a function of the current and time explicitly.

This explicit dependence on time should not be confused with the actual dependence on time of the voltage, part of which is caused by $R(t)$, and part by the actual time dependence of the current.

In Sec. 3.3 we prove frequency-power formulas of Type II by use of this dissipation function. In Sec. 3.4, however, in which we prove the formulas of Type III, we do not require the existence of such a state function, but instead only the functional relations of Eq. 3.5 and the fact that the matrix with i-k entry,

$$\frac{\partial f_i}{\partial v_k} \qquad (3.7)$$

is positive definite [43, Sec. 1.17].

Reactive, or energy storage, devices, on the other hand, do not have constitutive relations of the form of Eq. 3.5. For example, the instantaneous voltage of a capacitor is not a function of the current, but instead of the charge on the capacitor plates, which is the integral of the current. We discuss reactive devices by defining a new set of variables, x_i, as the integral of the v_i variables. The intent is that the f_i variables should, for reactive devices, be a function of the x_i variables:

$$f_i = f_i(x_k) \qquad (3.8)$$

Some devices with constitutive relations of this sort are (besides the capacitor): the inductor, with current i a function of the flux linkage λ, whose derivative is the voltage e; a mechanical spring, for which the force f is a function of the displacement x, whose time derivative is the velocity v; and an electromagnetic medium in which $\overline{E}$ is some function of the displacement vector $\overline{D}$, whose time-derivative is related (through one of Maxwell's equations) to $\overline{H}$, and in which $\overline{H}$ is some function of the magnetic induction vector $\overline{B}$, whose time-derivative is related to $\overline{E}$.

If these constitutive relations are "reciprocal" in the sense that

$$\frac{\partial f_i(x_k)}{\partial x_\ell} = \frac{\partial f_\ell(x_k)}{\partial x_i} \tag{3.9}$$

then the line integral in multidimensional x_i space

$$U(x_i) = \Sigma_i \int f_i dx_i \tag{3.10}$$

defines a state function [42, Sec. 6.10; 80], independent of the path of integration, where

$$f_i = \frac{\partial U}{\partial x_i} \tag{3.11}$$

Defined in this way, the function U is the energy, since its time derivative

$$\frac{dU}{dt} = \Sigma_i \frac{\partial U}{\partial x_i} \frac{dx_i}{dt} = \Sigma_i f_i v_i \tag{3.12}$$

is the power input to the device.

The treatment of time-varying reactive systems is similar. The constitutive relations of Eq. 3.8 are replaced by ones with explicit time dependence

$$f_i = f_i(x_k, t) \tag{3.13}$$

If Eq. 3.9 holds, then the energy function defined in Eq. 3.10 has an explicit time dependence, and Eq. 3.12 does not hold. The device cannot, therefore, be called lossless, since there can be net generation or destruction of energy within the framework of the variables considered. This is discussed further in Sec. 4.4, in connection with the time-varying capacitor.

Other energy state functions can be derived from U by Legendre transformations [30, Sec. 7.1], and in Sec. 3.2 we show how the existence of this U function, or any energy state function related to it by a Legendre transformation, leads to frequency power formulas of Type I, the Manley-Rowe formulas. Energy state functions that are not related to U by a Legendre transformation also lead to frequency-power formulas, but the interpretation of the expressions is not always clear.

In Sec. 3.5 frequency-power formulas of Type IV are proved, for reactive devices that obey constitutive relations of the form of Eq. 3.13, provided that the matrix with i-k entry

$$\frac{\partial f_i}{\partial x_k} \tag{3.14}$$

is positive definite [43, Sec. 1.17]. We do not use the state function U.

In summary, the important points to remember about this notation for variables are that products of f_i and v_i give power flow or power input, and that the x_i variables are the time-integrals of the v_i.

The Frequencies. For generality we do not wish to use only frequencies of the form $m\omega_1 + n\omega_0$. Perhaps in practice these frequency constraints are sufficient; our analysis is nevertheless more general and simpler, because we choose not to specify what frequencies are present. We merely assume that there are a number of frequencies present, and that only a few of these are independent. The results reduce to the formulas given earlier when we specify that the frequencies are to be of the form $m\omega_1 + n\omega_0$ with ω_1 and ω_0 independent. They also reduce to other formulas given before [18, 99, 12, 120] when the appropriate frequencies are specified.

We assume that the device is operating in the steady state, so that each of the variables can be written in a multiply-periodic Fourier series

$$v_i(t) = v_{i0} + \mathrm{Re}\,\Sigma_\alpha (v_i)_\alpha e^{j\omega_\alpha t} \tag{3.15}$$

$$f_i(t) = f_{i0} + \mathrm{Re}\,\Sigma_\alpha (f_i)_\alpha e^{j\omega_\alpha t} \tag{3.16}$$

so that

$$x_i(t) = x_{i0} + v_{i0}t + \mathrm{Re}\,\Sigma_\alpha (x_i)_\alpha e^{j\omega_\alpha t} \tag{3.17}$$

where the frequencies of the system are ω_α, α being an index. Because each v_i is the time derivative of x_i, their Fourier coefficients are related:

$$(v_i)_\alpha = j\omega_\alpha (x_i)_\alpha \tag{3.18}$$

Because the operation is in the steady state, the state functions (if defined) are bounded and can be represented in the same form:

$$G(t) = G_0 + \mathrm{Re}\,\Sigma_\alpha G_\alpha e^{j\omega_\alpha t} \tag{3.19}$$

and

$$U(t) = U_0 + \mathrm{Re}\,\Sigma_\alpha U_\alpha e^{j\omega_\alpha t} \tag{3.20}$$

The set of frequencies ω_α is large enough to include all frequencies needed for Eqs. 3.15, 3.16, 3.17, 3.19, and 3.20. In these equations, the summations over α are not intended to include a given frequency and its negative, and are not intended to include the d-c terms, which are shown separately. Thus, if some frequency ω is included in the summation, $-\omega$ is not. One way to insure this requirement is to sum over only positive frequencies, although this method is not the only way, nor is it always the best. We will henceforth assume, in all summations of this sort, that we do not sum over both ω and $-\omega$, but only over one of them (either one).

If a negative frequency is included in the summation, then the Fourier coefficients associated with this negative frequency are the complex conjugates of the Fourier coefficients for the corresponding positive frequency. In addition, the real power associated with a negative frequency is the same as the real power associated with the corresponding positive frequency, whereas the reactive power associated with a negative frequency is minus the reactive power associated with the corresponding positive frequency.

In general, the systems we discuss are time-varying. The explicit time dependence of the variables arises through specific functions of time; these functions can, in the steady state, be represented in terms of some frequencies ω_β, where β is an index. The set of frequencies $\{\omega_\alpha\}$, which includes all frequencies necessary for the expansions of Eqs. 3.15, 3.16, 3.17, 3.19, and 3.20, thus probably includes many of the ω_β as well as many other frequencies.

Not all the ω_α frequencies are independent; there are constraints that enable us to define a set of independent frequencies, in terms of which all ω_α can be expressed. For example, if the frequencies ω_α are of the form $m\omega_1 + n\omega_0$, we may define as independent frequencies ω_1 and ω_0, although this is not the only possible choice.

In general, any particular frequency ω_α depends on some or all of the independent frequencies. The frequencies ω_β, which go towa specifying the explicit time dependence of the system, thus depend on some of the independent frequencies, possibly not all. How many of the independent frequencies the ω_β depend on is to some extent determined by the independent frequencies and how they are chosen. We have considerable freedom in choosing them; thus, with one choice we might find that the ω_β frequencies depend on more of the independent frequencies than with another choice. Because there are frequency-power formulas associated with each independent frequenc on which no ω_β depend, it is wise to choose the independent frequencies in such a way that only one (or only a few) is necessary to specify all the ω_β.

An example may clarify this. Suppose a capacitor is varied mech- anically at frequency ω_p and its harmonics, and that a signal of frequency ω_s is put on the electrical terminals. Sidebands will be generated with frequencies of the form $\omega_s + n\omega_p$, where n is a positive or negative integer. Considering only the electrical variables, the device is represented by a time-varying capacitance, with power exchange at frequencies $n\omega_p + \omega_s$. There are two independent frequencies, and we might choose for them ω_s and $\omega_s + \omega_p$. But if we do, the frequency that specifies the explicit time dependence of the system, ω_p, depends on <u>both</u> independent frequencies, and there are no frequency-power formulas. On the other hand, if we choose as independent frequencies ω_p and $\omega_p + \omega_s$, there is a frequency-power formula associated with the independent frequency $\omega_p + \omega_s$.

Let us call ω_a, where a is an index, the independent frequencies on which the ω_β, which specify the explicit time dependence of the system, do <u>not</u> depend. Then all partial derivatives of the form

$$\frac{\partial \omega_\alpha}{\partial \omega_a} \tag{3.21}$$

are well defined, and all partial derivatives of the form

$$\frac{\partial \omega_\beta}{\partial \omega_a} \tag{3.22}$$

vanish. There is a frequency-power formula associated with each ω_a.

To review this notation:

(1) Frequencies ω_α are all frequencies present in the Fourier analysis of the <u>actual</u> time dependences of the variables.

(2) Frequencies ω_β are all frequencies that help to specify the <u>explicit</u> time dependence of the constitutive relations.

(3) Frequencies ω_a are all independent frequencies on which the ω_β do not depend.

For simplicity, in the proofs of the formulas we use the phases $\phi_\alpha = \omega_\alpha t$, $\phi_\beta = \omega_\beta t$, and $\phi_a = \omega_a t$. The phases ϕ_β do not depend on the ϕ_a, and any finite change θ_a in an independent phase ϕ_a induces changes

$$\theta_\alpha = \frac{\partial \omega_\alpha}{\partial \omega_a} \theta_a$$

in the dependent phases ϕ_α.

With the variables given in the form of Eqs. 3.15 and 3.16, the time-average power input P can be written as

$$P = \Sigma_i P_{io} + \Sigma_\alpha P_\alpha \tag{3.23}$$

where

$$P_{io} = v_{io} f_{io} \tag{3.24}$$

is the d-c power input caused by the i-th variables, and where

$$P_\alpha = \tfrac{1}{2}\mathrm{Re}\,\Sigma_i\,(f_i)^*_\alpha(v_i)_\alpha \tag{3.25}$$

is the total power input at frequency ω_α. Frequency-power formulas of Types I and III express relations among the P_{i0} and P_α.

Similarly, the reactive power at each frequency ω_α is generally

$$Q_\alpha = \tfrac{1}{2}\,\mathrm{Im}\,\Sigma_i(f_i)^*_\alpha(v_i)_\alpha \tag{3.26}$$

or its negative. Frequency-power formulas of Types II and IV express relations among the Q_α.

There are four difficulties with interpreting Eq. 3.26 as reactive power that do not arise in the interpretation of Eq. 3.25 as real power at frequency ω_α. First, there are in use two definitions of reactive power, each the negative of the other. Guillemin [33, Chap. 7, Sec. 4] defines the "vector power"

$$P + jQ = \tfrac{1}{2}\,E^*I \tag{3.27}$$

where E and I are peak amplitudes of voltage and current respectively. According to this definition, reactive power flows out of inductors and into capacitors. This is the reverse of the customary definition, used, for example, by Manley and Rowe [60]

$$P + jQ = \tfrac{1}{2}\,EI^* \tag{3.28}$$

According to this convention, which is the one we use in later chapters, reactive power flows out of capacitors and into inductors. Each scheme is self-consistent, and it is an arbitrary choice that must be made.

The second difficulty with interpreting the Q_α of Eq. 3.26 as reactive power is that we have not specified what the variables f and v stand for. If v stands for the current and f for the voltage, then Q_α is the negative of the reactive power as we use it, whereas if v stands for the voltage and f for the current, then Q_α is just equal to the reactive power as we define it.

The third difficulty is that with multiport devices, the contribution to Q_α at one port might be the ordinary reactive power there, whereas the contribution at another port might be the negative of the reactive power. This happens, for example, in a gyrator. In this case, we cannot interpret the Q_α of Eq. 3.26 as either the reactive power or minus the reactive power.

The fourth difficulty occurs if some of the frequencies ω_α are negative. Then the Q_α associated with these negative frequencies are just the negatives of the Q_α that would be associated with the corresponding positive frequencies ω_α.

These four points must be kept in mind in interpreting frequency-power formulas of Types II and IV.

Using the notation developed in this section, we can easily prove the four types of frequency-power formulas.

3.2 Type I. Manley-Rowe Formulas

Here are the purely mathematical steps leading to the Manley-Rowe formulas. We use the notation explained in Sec. 3.1, and start with an energy state function.

We assume we have a "reactive" device: more precisely, we assume that the device has a time-varying energy state function

$$U = U(x_i, t) \tag{3.29}$$

with variables

$$f_k(x_i, t) = \frac{\partial U(x_i, t)}{\partial x_k} \tag{3.30}$$

Using the frequency notation of Sec. 3.1(The Frequencies), we write all f_i, x_i, and U in the form of Eqs. 3.16, 3.17, and 3.20. In our applications the v_{i0} quantities, which contribute toward d-c power input, are in the nature of frequencies, because they appear multiplied by t. (For example, the position of a rotating shaft would be in the form of Eq. 3.17 with v_{i0} the average speed). We assume that these v_{i0} are, for the purpose of defining independent frequencies, included among the ω_α frequencies. Thus all quantities of the form

$$\frac{\partial v_{i0}}{\partial \omega_a} \tag{3.31}$$

are well defined.

Now let us make a small change $\delta\phi_a$ in one of the independent phases, ϕ_a, keeping all the others fixed. We can compute, to first order, the change δU in the state function U. We find from Eq. 3.20

$$\delta U = \operatorname{Re} \Sigma_\alpha jU_\alpha \frac{\partial \omega_\alpha}{\partial \omega_a} e^{j\omega_\alpha t} \delta\phi_a \tag{3.32}$$

We have another way of calculating δU. Because the explicit time dependence of U does not change (remember none of the ϕ_β depend on ϕ_a), δU can be expressed in terms of its dependence on the x_i variables as

$$\delta U = \Sigma_i f_i \delta x_i \tag{3.33}$$

which becomes, from Eqs. 3.16 and 3.17

$$\delta U = \Sigma_i \left[f_{i0} + \operatorname{Re} \Sigma_\alpha (f_i)_\alpha e^{j\omega_\alpha t} \right] \times \left[\frac{\partial v_{i0}}{\partial \omega_a} \delta\phi_a + \operatorname{Re} \Sigma_\alpha j(x_i)_\alpha \frac{\partial \omega_\alpha}{\partial \omega_a} e^{j\omega_\alpha t} \delta\phi_a \right] \tag{3.34}$$

These two calculations of δU must agree for all time. Equation 3.32 shows δU as a sum of sinusoids, which vanishes, when averaged with respect to time. Applying the same test to Eq. 3.34, we find that its average value is

$$\Sigma_i \left[f_{io} \frac{\partial v_{io}}{\partial \omega_a} \right] \delta\phi_a + \tfrac{1}{2} \Sigma_i \Sigma_\alpha \frac{\partial \omega_\alpha}{\partial \omega_a} \mathrm{Re} \left[j(x_i)_\alpha (f_i)_\alpha{}^* \right] \delta\phi_a \tag{3.35}$$

which must vanish. Thus, by use of Eqs. 3.24 and 3.25

$$\Sigma_i \frac{P_{io}}{v_{io}} \frac{\partial v_{io}}{\partial \omega_a} + \Sigma_\alpha \frac{P_\alpha}{\omega_\alpha} \frac{\partial \omega_\alpha}{\partial \omega_a} = 0 \tag{3.36}$$

for each frequency ω_a. These are the Manley-Rowe formulas, extended to include d-c power input and more general frequency constraints. They reduce to the ordinary form of the Manley-Rowe formulas, Eqs. 2.2 and 2.3, when the d-c powers vanish, when the device considered is the ordinary nonlinear capacitor, and when the frequencies of the system are in the form $m\omega_1 + n\omega_0$, with ω_1 and ω_0 chosen as independent frequencies.

The d-c terms in Eq. 3.36 have proved useful so far only for rotating machines, so in quoting these formulas in later chapters the d-c terms are usually omitted. In Chapters 4 and 6 we discuss several physical systems that have energy state functions, and therefore obey these formulas.

3.3 Type II

Frequency-power formulas of Type II relate reactive power in nonlinear dissipative devices. We prove them here with a state function argument similar to that used for proving the Manley-Rowe formulas in Sec. 3.2.

We assume the device is "dissipative:" more precisely, we assume that it has a time-varying dissipation function

$$G(v_i,\ t) \tag{3.37}$$

such that

$$f_k(v_i,\ t) = \frac{\partial G(v_i,\ t)}{\partial v_k} \tag{3.38}$$

Using the frequency notation of Sec. 3.1 (The Frequencies), we write all v_i, f_i, and G in the form of Eqs. 3.15, 3.16, and 3.19.

Now let us make a small change $\delta\phi_a$ in one of the independent phases ϕ_a, keeping all the others fixed. We can compute, to first order, the change δG in the state function G. We find from Eq. 3.19

$$\delta G = \text{Re}\, \Sigma_\alpha j G_\alpha \frac{\partial \omega_\alpha}{\partial \omega_a} e^{j\omega_\alpha t} \delta\phi_a \tag{3.39}$$

Notice that δG consists of sinusoids, and therefore its time-average value vanishes.

But we have another way of calculating δG. Because the explicit time dependence of G does not change (remember that none of the ϕ_β depends on ϕ_a), δG can be expressed in terms of its dependence on the v_i variables as

$$\delta G = \Sigma_i f_i \delta v_i \tag{3.40}$$

which becomes, from Eqs. 3.15 and 3.16

$$\delta G = \Sigma_i \left[f_{i0} + \text{Re}\, \Sigma_\alpha (f_i)_\alpha e^{j\omega_\alpha t} \right] \left[\text{Re}\, \Sigma_\alpha j (v_i)_\alpha \frac{\partial \omega_\alpha}{\partial \omega_a} e^{j\omega_\alpha t} \delta\phi_a \right] \tag{3.41}$$

Because the two calculations for δG must agree for all time, the average value of Eq. 3.41

$$\Sigma_i \Sigma_\alpha \tfrac{1}{2} \text{Re}\, j (v_i)_\alpha (f_i)_\alpha^* \frac{\partial \omega_\alpha}{\partial \omega_a} \delta\phi_a \tag{3.42}$$

must vanish, and so by use of Eq. 3.26

$$\Sigma_\alpha \frac{\partial \omega_\alpha}{\partial \omega_a} Q_\alpha = 0 \tag{3.43}$$

for each frequency ω_a. These are the frequency-power formulas of Type II; some devices that obey them are discussed in Chapter 5.

In interpreting these formulas, the remarks following Eq. 3.26 must be kept in mind. In particular, the gyrator does not obey Eq. 3.43, even though one can define a dissipation function G for it; this is because the Q_α that result are neither the reactive powers nor their negatives.

3.4 Type III

Page [74, 75] and Pantell [77] have given frequency-power formulas relating real power in positive nonlinear resistors. By "positive," it is meant that the incremental resistance

$$R_i = \frac{de}{di} \tag{3.44}$$

is nonnegative.

It is interesting to note that both Page's and Pantell's forms of the formulas have advantages. Pantell's results are easily applied to devices with more than one independent frequency, whereas Page's formulas are not written directly in terms of the independent frequencies. On the other hand, when they both apply (as they do, for example, to harmonic generators) Page's formulas are more powerful. Here we show the logical generalization of each — we retain all the power of Page's formulas, but express them in terms of dependent and independent frequencies to make them easier to apply. The method of proof used is a modification of Page's method.

We assume the device considered is "dissipative," in that time-varying constitutive relations relate the f_i and v_i variables in the form of Eq. 3.5. Using the representations of Eqs. 3.15 and 3.16, we compute the change in the variables caused by a finite change θ_a in one of the independent phases ϕ_a. This produces a finite change

$$\theta_\alpha = \frac{\partial\omega_\alpha}{\partial\omega_a}\theta_a \tag{3.45}$$

in each dependent phase ϕ_α, so the change in each v_i is

$$\Delta v_i = \text{Re}\ \Sigma_\alpha (v_i)_\alpha e^{j\omega_\alpha t}(e^{j\theta_\alpha} - 1) \tag{3.46}$$

and

$$\Delta f_i = \text{Re}\ \Sigma_\alpha (f_i)_\alpha e^{j\omega_\alpha t}(e^{j\theta_\alpha} - 1) \tag{3.47}$$

If we multiply Δv_i by Δf_i and sum over i, and then take the time average of the sum, we find

$$\begin{aligned} h_a(\theta_a) &= \left\langle \Sigma_i \Delta v_i \Delta f_i \right\rangle \\ &= \tfrac{1}{2}\text{Re}\ \Sigma_i \Sigma_\alpha (v_i)_\alpha (f_i)_\alpha^*(1 - \cos\theta_\alpha) \end{aligned} \tag{3.48}$$

and so the quantity h_a defined in Eq. 3.48 can be written

$$h_a(\theta_a) = \Sigma_\alpha P_\alpha (1 - \cos\theta_\alpha) \tag{3.49}$$

by use of Eq. 3.25.

So far we have not proved anything of value, for there is no estimate of the quantity $h_a(\theta_a)$. However, because the f_i are functions of the v_i variables, the changes Δf_i can be expressed in terms of line integrals in v_i-space:

$$\Delta f_i = \Sigma_k \int_{v_k}^{v_k + \Delta v_k} \frac{\partial f_i}{\partial v_k} dv_k \tag{3.50}$$

Thus $\Sigma_i \Delta f_i \Delta v_i$, and hence its time-average $h_a(\theta_a)$, is nonnegative if the matrix with i-k entry

$$\frac{\partial f_i}{\partial v_k} \tag{3.51}$$

is positive definite (or positive semidefinite). Equation 3.51 reduces to the previous criterion of positiveness, Eq. 3.44, when there is only one set of variables; it is the logical generalization for devices with more variables.

Assuming that the matrix of Eq. 3.51 is positive definite (or semi-definite), then frequency-power formulas of Type III are written

$$h_a(\theta_a) = \Sigma_\alpha P_\alpha (1 - \cos \frac{\partial \omega_\alpha}{\partial \omega_a} \theta_a) \geq 0 \tag{3.52}$$

for all values of θ_a; these are relationships of some value, because sums of powers are constrained to be positive.

The requirement that the matrix of Eq. 3.51 be positive definite has been termed "local passivity" by Duinker [19, Ch. II]. Locally passive devices exhibit, at each operating point, small-signal equations that define a passive system. An example of a device that is not locally passive at all points is the tunnel diode [22, 97], with a volt-ampere characteristic such as that of Fig. 3.1. The points of

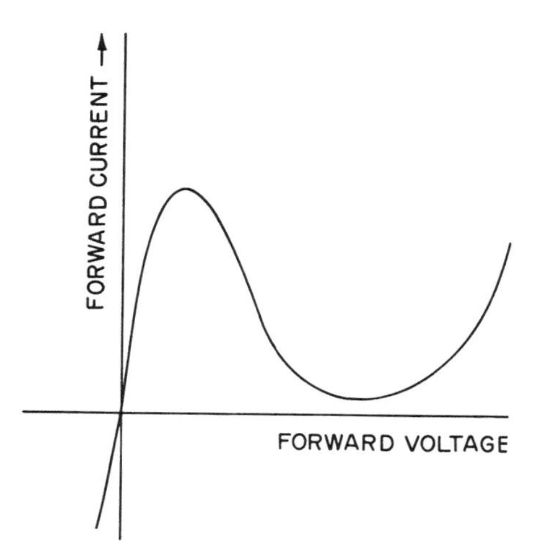

Fig. 3.1. Tunnel diode d-c volt-ampere curve. The negative-resistance region is not a region of local passivity

negative slope are not points of local passivity; the small-signal equivalent circuit at such points is active, not passive.

Equations 3.52 are the basic frequency-power formulas of Type

III; they are in a form similar to Page's relations [74]. Pantell's form of the formulas are obtained in the limit as $\theta_a \to 0$:

$$\Sigma_\alpha P_\alpha \left(\frac{\partial \omega_\alpha}{\partial \omega_a} \right)^2 \geq 0 \tag{3.53}$$

These are sometimes more easily applied to practical problems than Eqs. 3.52, but they are less powerful.

It is interesting to note that even more general formulas can be written. If in the preceding proof we make a finite, but different, change in each independent phase ϕ_a, the finite change θ_α in each dependent phase ϕ_α is

$$\theta_\alpha = \Sigma_a \frac{\partial \omega_\alpha}{\partial \omega_a} \theta_a \tag{3.54}$$

and the result of Eq. 3.52 becomes

$$\Sigma_\alpha P_\alpha \left[1 - \cos\left(\Sigma_a \frac{\partial \omega_\alpha}{\partial \omega_a} \theta_a \right) \right] \geq 0 \tag{3.55}$$

for all values of the parameters θ_a. Setting all θ_a equal to zero except one reduces this result to Eq. 3.52. In the limit as all $\theta_a \to 0$, but remain proportional to each other,

$$\Sigma_\alpha P_\alpha \left(\Sigma_a \frac{\partial \omega_\alpha}{\partial \omega_a} \chi_a \right)^2 \geq 0 \tag{3.56}$$

for all values of the parameters χ_a. Setting each χ_a in turn to one and all the others to zero, we obtain Eqs. 3.53.

The formulas can be generalized even more. Following the extension of Page [74], we let $\xi(x)$ be any nonincreasing positive function defined from 0 to ∞, and we define

$$g_\alpha(\theta_a) = \left(\Sigma_a \frac{\partial \omega_\alpha}{\partial \omega_a} \theta_a \right) \int_0^\infty \xi(x) \sin\left(\Sigma_a \frac{\partial \omega_\alpha}{\partial \omega_a} \theta_a x \right) dx \tag{3.57}$$

Then

$$\Sigma_\alpha P_\alpha g_\alpha(\theta_a) \geq 0 \tag{3.58}$$

for all values of the parameters θ_a. If $\xi(x)$ is the function

$$\xi(x) = \begin{cases} 1 & 0 \leq x < 1 \\ 0 & x \geq 1 \end{cases} \tag{3.59}$$

then Eq. 3.58 reduces to Eq. 3.55. The proof of Eq. 3.58 is identical to the proof given by Page of a similar result [74].

Frequency-power formulas of Type III apply to time-varying locally-passive dissipative systems, some examples of which are discussed in Chapter 5. In later chapters we usually quote the formulas as given by Pantell, Eqs. 3.53, with the understanding that the other forms, Eqs. 3.52, 3.55, 3.56, and 3.58, are also valid.

3.5 Type IV

Frequency-power formulas of Type IV relate reactive power in reactances. Like the formulas of Type III, they are inequalities, instead of equalities. They are proved in a similar way and require a similar condition of local passivity.

When interpreting the formulas, one must keep in mind the points discussed immediately following Eq. 3.26. For inductive devices the formulas will be written (using the actual reactive powers) with a $\geq$ sign; and for capacitive devices the formulas ought to hold with a $\leq$ sign. Multiport devices with both electric and magnetic energy storage do not obey the formulas, because of the ambiguity of sign, and likewise distributed systems do not obey them. Their main use is with single nonlinear time-varying elements, or with configurations of similar-type elements, with coupling (such as a rotating machine).

When the formulas are quoted in future chapters, the form for inductive-type devices is given, unless otherwise stated. It is to be understood that if the device is actually capacitive, the sign should be reversed. In some places we give both forms for convenience.

To prove the formulas, we assume the device is "reactive:" specifically, we assume it has constitutive relations of the form of Eq. 3.13, between the f_i and x_i variables. Using the representations of Eqs. 3.16 and 3.17, we compute the finite change in the variables caused by a finite change θ_a in one of the independent phases ϕ_a. As in Sec. 3.4, this produces a finite change

$$\theta_\alpha = \frac{\partial \omega_\alpha}{\partial \omega_a} \theta_a \tag{3.60}$$

in each dependent phase ϕ_α, and so the change in each of the variables is

$$\Delta f_i = \operatorname{Re} \Sigma_\alpha (f_i)_\alpha e^{j\omega_\alpha t} (e^{j\theta_\alpha} - 1) \tag{3.61}$$

and

$$\Delta x_i = \frac{\partial v_{io}}{\partial \omega_a} \theta_a + \operatorname{Re} \Sigma_\alpha (x_i)_\alpha e^{j\omega_\alpha t} (e^{j\theta_\alpha} - 1) \tag{3.62}$$

If we multiply Δf_i by Δx_i and sum over i, and then take the time-average of the sum, we obtain

$$h_a'(\theta_a) = \left\langle \Sigma_i \Delta x_i \Delta f_i \right\rangle$$

$$= \tfrac{1}{2}\mathrm{Re}\ \Sigma_i \Sigma_\alpha (x_i)_\alpha (f_i)_\alpha^* (1 - \cos\theta_\alpha) \tag{3.63}$$

or

$$h_a'(\theta_a) = \Sigma_\alpha \frac{Q_\alpha}{\omega_\alpha} (1 - \cos\theta_\alpha) \tag{3.64}$$

where Q_α is given by Eq. 3.26.

What we have so far is not of much value, because there is no estimate of the quantity $h_a'(\theta_a)$. However, because the f_i variables are functions of the x_i variables, the changes Δf_i can be evaluated by line-integrals in x_i-space:

$$\Delta f_i = \Sigma_k \int_{x_k}^{x_k + \Delta x_k} \frac{\partial f_i}{\partial x_k}\, dx_k \tag{3.65}$$

Thus $\Sigma_i \Delta f_i \Delta x_i$, and hence its time-average $h_a'(\theta_a)$, is nonnegative if the matrix with i-k entry

$$\frac{\partial f_i}{\partial x_k} \tag{3.66}$$

is positive definite (or positive semidefinite). This condition is one of local passivity, just like the condition for dissipative devices in Sec. 3.4.

For locally-passive time-varying nonlinear reactances, then, frequency-power formulas of Type IV hold in the form

$$h_a'(\theta_a) = \Sigma_\alpha \frac{Q_\alpha}{\omega_\alpha} \left(1 - \cos \frac{\partial \omega_\alpha}{\partial \omega_a} \theta_a\right) \geq 0 \tag{3.67}$$

for all θ_a, provided care is used in interpreting the Q_α quantities, as discussed earlier.

Just as in Sec. 3.4, we can write other forms of these formulas. In the limit as $\theta_a \to 0$, we find

$$\Sigma_\alpha \frac{Q_\alpha}{\omega_\alpha} \left(\frac{\partial \omega_\alpha}{\partial \omega_a}\right)^2 \geq 0 \tag{3.68}$$

These formulas are easily applied to practical problems, although they are not as powerful as Eqs. 3.67. If in the preceding proof we make a finite change in each of the independent frequencies, we obtain frequency-power formulas similar to Eqs. 3.55:

$$\Sigma_\alpha \frac{Q_\alpha}{\omega_\alpha}\left[1 - \cos\left(\Sigma_a \frac{\partial\omega_\alpha}{\partial\omega_a}\theta_a\right)\right] \geq 0 \qquad (3.69)$$

for all values of the parameters θ_a. Setting all θ_a to zero except one reduces Eq. 3.69 to the simpler form of Eq. 3.67. If all θ_a approach zero, but remain proportional to each other, we find

$$\Sigma_\alpha \frac{Q_\alpha}{\omega_\alpha}\left(\Sigma_a \frac{\partial\omega_\alpha}{\partial\omega_a}\chi_a\right)^2 \geq 0 \qquad (3.70)$$

for all values of the parameters χ_a. Setting each χ_a in turn to one and all others to zero, we obtain Eq. 3.68. And finally, if $g_\alpha(\theta_a)$ is defined, as before, by Eq. 3.57, then

$$\Sigma_\alpha \frac{Q_\alpha}{\omega_\alpha} g_\alpha(\theta_a) \geq 0 \qquad (3.71)$$

for all values of the parameters θ_a.

In later chapters we usually quote the formulas of Type IV as given by Eq. 3.68, with the understanding that the other forms, Eqs. 3.67, 3.69, 3.70, and 3.71, are also valid.

There are devices that are not locally passive, for which we would like to have frequency-power formulas of Type IV. One such device is the induction motor, in which the electrical variables for each fixed shaft position form a locally passive system. When the mechanical variable is added, however, there can be regions of unstable equilibrium, and regions in which the electromechanical system is not locally passive. Can we find frequency-power formulas of Type IV for similar cases?

In some cases we can. If a small matrix within the matrix of Eq. 3.66 is positive definite, let us call the remaining variables (those that are not related by this submatrix) the "offending" variables. Thus, for the induction motor, the mechanical variables would be the offending ones. It may happen that none of the offending variables depends in its actual time variation on one or more of the ω_a independent variables. In this case, the corresponding h_a' is nonnegative, since the summation in its definition of Eq. 3.63 can be restricted to the non-offending variables. Thus, with this "weak positive-definite" system, there are still some frequency-power formulas applicable.

PART II

SPECIFIC SYSTEMS

Chapter 4

LUMPED REACTIVE SYSTEMS

In this chapter we discuss some lumped devices that obey frequency-power formulas of Types I and IV. Generally speaking, these are reactive systems, although in Sec. 4.11 it is observed that linear resistors obey formulas of Type IV. Our purpose is to establish a partial list of devices that obey the formulas.

In Sec. 4.1 we discuss the energy functions that are suitable for proving the Manley-Rowe formulas, and show that one may use the internal energy (expressed in terms of the proper variables), or any energy function related to it by a Legendre transformation.

The remaining sections deal with specific devices. Sections 4.2, 4.3, and 4.5 discuss nonlinear energy storage elements — capacitors, inductors, and mechanical elements. In Sec. 4.4, we show that a time-varying capacitor, although not lossless, obeys the Manley-Rowe formulas.

Many rotating machines (those without commutators) obey the formulas, and in Sec. 4.6 a general model for such a machine is discussed. A logical generalization of this model is discussed in Sec. 4.7; it includes both magnetic and electric coupling to mechanical circuits, and obeys the Manley-Rowe formulas.

There are very few idealized nonlinear circuit elements. The ideal diode is one; others, called "traditors," have been proposed by Duinker [20]. In Sec. 4.8 we show that these, unlike the ideal diode, obey the Manley-Rowe formulas.

Some degenerate elements that satisfy frequency-power formulas of Types I or IV are discussed in Sec. 4.9. These devices generally satisfy the formulas in a trivial way, with all P_α or Q_α (or both) equal to zero. The major purpose of discussing them at all is to aid the analysis of entire networks. It is shown in Sec. 4.10 that a network of elements, each of which obeys frequency-power formulas of Type I, itself obeys the formulas. Thus, in particular, these "degenerate" elements of Sec. 4.9 can appear in such networks. Included are the ideal transformer, the gyrator, the differential gear, and (in a formal way) the maser.

Many distributed systems obey the formulas of Type I; these are discussed in Chapter 6.

4.1 Energy Functions

It is shown in Sec. 3.2 that any device with an energy state function obeys the Manley-Rowe formulas. The function used there is (at least in the time-invariant case) the energy of the device, expressed in terms of the x_i variables, and the expression for P_a

$$P_a = \tfrac{1}{2}\mathrm{Re}\,\Sigma_i (f_i)_a^{*}(v_i)_a \tag{4.1}$$

is easily interpreted as the power input to the device at frequency ω_a.

The same method, however, applies to any energy function, as a function of any of the variables. The formulas are of the same form, but the expression for P_a is different from that of Eq. 4.1. Since a given physical system can be described by many different energy functions, it obeys many different Manley-Rowe formulas. If we want to interpret the P_a as power inputs at frequency ω_a, only a few energy functions are appropriate. What combinations of energy functions and variables are suitable?

It is known that the internal energy U, as a function of the x_i variables, is suitable; we show here that any function related to $U(x_i, t)$ by a Legendre transformation [30, Sec. 7.1] can be used.

Because

$$f_i = \frac{\partial U}{\partial x_i} \tag{4.2}$$

the most general form for any energy function related to $U(x_i, t)$ by a Legendre transformation is

$$U'(x_\ell, f_k, t) = U(x_i, t) - \Sigma_k x_k f_k \tag{4.3}$$

or its negative, where i is an index that runs over all the variables, k is an index that runs over some of the variables, and ℓ is an index that runs over the rest. The Legendre transformation of Eq. 4.3 defines explicitly the variables on which the new function depends and

$$\frac{\partial U'}{\partial x_\ell} = f_\ell \tag{4.4}$$

and

$$\frac{\partial U'}{\partial f_k} = -x_k \tag{4.5}$$

To have U' satisfy the condition that it be bounded, we must require that all v_{k0} vanish, as is evident from Eqs. 3.17 and 4.3.

The procedure of Sec. 3.2, when applied to $U'(x_\ell, f_k, t)$ yields Manley-Rowe formulas in the form of Eq. 3.36 with

$$P_{k0} = 0 \tag{4.6}$$

$$P_{\ell 0} = v_{\ell 0} f_{\ell 0} \tag{4.7}$$

and

$$\begin{aligned} P_\alpha &= \tfrac{1}{2}\mathrm{Re}\left[\Sigma_k j\omega_\alpha \left(\frac{\partial U'}{\partial f_k}\right)_\alpha^* (f_k)_\alpha + \Sigma_\ell j\omega_\alpha \left(\frac{\partial U'}{\partial x_\ell}\right)_\alpha^* (x_\ell)_\alpha\right] \\ &= \tfrac{1}{2}\mathrm{Re}\left[-\Sigma_k j\omega_\alpha (x_k)_\alpha^* (f_k)_\alpha + \Sigma_\ell j\omega_\alpha (f_\ell)_\alpha^* (x_\ell)_\alpha\right] \\ &= \tfrac{1}{2}\mathrm{Re}\left[\Sigma_k (v_k)_\alpha^* (f_k)_\alpha + \Sigma_\ell (v_\ell)_\alpha (f_\ell)_\alpha^*\right] \\ &= \tfrac{1}{2}\mathrm{Re}\ \Sigma_i (f_i)_\alpha^* (v_i)_\alpha \end{aligned} \tag{4.8}$$

These are exactly the forms obtained by using $U(x_i, t)$, so any interpretation given the P_α in that case applies here also.

It is interesting that a given physical system obeys Manley-Rowe formulas with many different forms for P_α. One can even show that for any (reasonable) function of the voltage $w(e)$ there is a set of Manley-Rowe formulas with

$$P_\alpha = \tfrac{1}{2}\mathrm{Re}\ w_\alpha \left(\frac{ei}{w}\right)_\alpha^* \tag{4.9}$$

that hold for the nonlinear capacitor. By thinking up different energy functions and different variables for them to depend on, we come upon many other forms. Because our desire is to interpret the P_α as power inputs at frequency ω_α, we will not pursue these other forms for P_α further. We find it sufficient to use $U(x_i, t)$ or a function related to it by a Legendre transformation.

In a similar way, a system that obeys frequency-power formulas of any type also obeys similar formulas with different P_α or Q_α. Since our purpose is to interpret the P_α and Q_α as real and reactive power at frequency ω_α, we will not have occasion to discuss these other forms.

4.2 Nonlinear Capacitor

This device is probably of most practical importance; it is the one originally discussed by Manley and Rowe [60]. A nonlinear capacitor without losses and without dispersion is characterized by a single-valued charge-voltage relation

$$e = e(q) \tag{4.10}$$

The internal energy $W_c(q)$ is

$$W_c(q) = \int_0^q e(q)\, dq \tag{4.11}$$

so that

$$e = \frac{dW_c(q)}{dq} \tag{4.12}$$

and the coenergy [14, 62] is defined by the Legendre transformation

$$W_c'(e) = eq - W_c(q) \tag{4.13}$$

These energy functions can be given a graphical interpretation, as shown on the voltage-charge characteristic in Fig. 4.1.

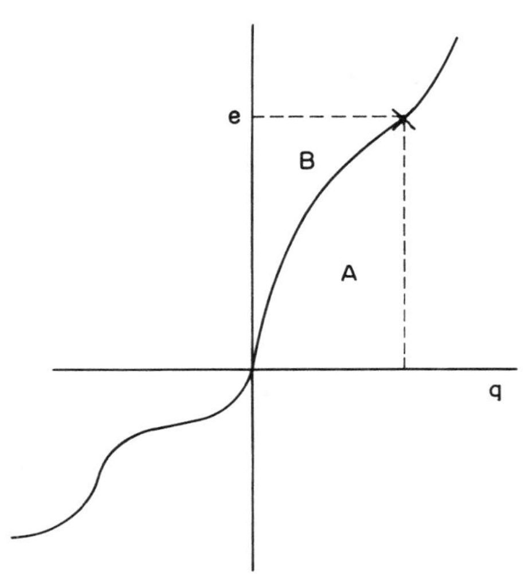

Fig. 4.1. A typical charge-voltage characteristic of a nonlinear capacitor. At the operating point indicated, the energy and coenergy are represented by the areas A and B respectively

Either the energy or the coenergy can be used to obtain frequency power formulas of the form (there is no d-c power)

$$\Sigma_\alpha \frac{P_\alpha}{\omega_\alpha} \frac{\partial \omega_\alpha}{\partial \omega_a} = 0 \tag{4.14}$$

with P_α given by

$$P_\alpha = \tfrac{1}{2}\mathrm{Re}\; e_\alpha i_\alpha^* \tag{4.15}$$

or simply the power into the capacitor at frequency ω_α.

The nonlinear capacitor also obeys the various forms of frequenc power formulas of Type IV discussed in Sec. 3.5, such as

$$\Sigma_a \left(\frac{\partial \omega_a}{\partial \omega_a} \right)^2 \frac{Q_a}{\omega_a} \leq 0 \tag{4.16}$$

where

$$Q_a = \tfrac{1}{2} \operatorname{Im} e_a i_a^* \tag{4.17}$$

is the reactive power into the capacitor at frequency ω_a.

Note that a linear capacitor satisfies these formulas in a trivial way, with all $P_a = 0$ and all $Q_a \leq 0$. The frequency-power formulas lose their importance, since much stronger conservation principles hold. Nevertheless, the frequency-power formulas are still valid, so the inclusion of linear capacitors in a network does not prevent it from obeying the formulas.

4.2 Nonlinear Inductor

The nonlinear inductor is treated similarly. It is characterized by a relation between the current and the flux linkage of the form

$$i = i(\lambda) \tag{4.18}$$

and so the stored magnetic energy is

$$W_m(\lambda) = \int_0^\lambda i(\lambda)\, d\lambda \tag{4.19}$$

with

$$i = \frac{dW_m(\lambda)}{d\lambda} \tag{4.20}$$

The use of $W_m(\lambda)$ as an energy state function leads to frequency-power formulas of the form of Eq. 4.14 with P_a given by Eq. 4.15. The magnetic coenergy $W_m'(i)$

$$W_m'(i) = i\lambda - W_m(\lambda) \tag{4.21}$$

can also be used, with the same result.

The constitutive relation of Eq. 4.18 leads to frequency-power formulas of Type IV of the form

$$\Sigma_a \left(\frac{\partial \omega_a}{\partial \omega_a} \right)^2 \frac{Q_a}{\omega_a} \geq 0 \tag{4.22}$$

or of any of the other forms derived in Sec. 3.5, with

$$Q_a = \tfrac{1}{2} \operatorname{Im} e_a i_a^* \tag{4.23}$$

equal to the reactive power input to the inductor. Notice that in the frequency-power formulas of Eq. 4.22 the sign is different from the corresponding formula for the nonlinear capacitor, Eq. 4.16.

Linear inductors satisfy Eqs. 4.14 and 4.22 in a trivial way, with all $P_\alpha = 0$ and $Q_\alpha \geq 0$. The frequency-power formulas lose their importance, but not their validity.

Two or more coils coupled nonlinearly also obey the formulas. Suppose, for two coils, the constitutive relations are written

$$\lambda_1 = \lambda_1(i_1, i_2) \tag{4.24}$$

and

$$\lambda_2 = \lambda_2(i_1, i_2) \tag{4.25}$$

Then formulas of Type IV hold in the form of Eq. 4.22, with

$$Q_\alpha = \tfrac{1}{2}\,\mathrm{Im}\left[(e_1)_\alpha (i_1)_\alpha^* + (e_2)_\alpha (i_2)_\alpha^*\right] \tag{4.26}$$

being the reactive power at frequency ω_α flowing into both coils.

Also, if

$$\frac{\partial \lambda_1(i_1, i_2)}{\partial i_2} = \frac{\partial \lambda_2(i_1, i_2)}{\partial i_1} \tag{4.27}$$

as is necessary for the device to be lossless, then the magnetic co-energy

$$W_m'(i_1, i_2) = \int \lambda_1\, di_1 + \int \lambda_2\, di_2 \tag{4.28}$$

is well defined [42, Sec. 6.10; 80]. This function leads to frequency-power formulas of Type I, in the form of Eq. 4.14, with

$$P_\alpha = \tfrac{1}{2}\mathrm{Re}\left[(e_1)_\alpha (i_1)_\alpha^* + (e_2)_\alpha (i_2)_\alpha^*\right] \tag{4.29}$$

interpreted as the power input to both coils at frequency ω_α. The extension to more than two coils is obvious.

4.4 Time-Varying Capacitor

This device probably causes more confusion than any other we discuss. The confusion centers around the meaning of the terms, "time-varying" and "lossless."

A given physical device may be accurately represented by any of several models, depending on circumstances. The time-varying capacitor is a model (not itself a device) that accurately describes many physical systems with some excitation previously assigned. Many people consider a variable capacitor (such as a radio tuning condenser) to be time-varying; our point of view is that if the mechanical motion is specified in advance, then a valid model, as far as the electrical variables are concerned, for this device is a time-varying capacitor. If the capacitance is linear, the equation of motion is

$$q(e, t) = C(t)\, e \tag{4.30}$$

where C(t) is the capacitance.

We obviously cannot predict, within the framework of the time-varying model, anything about the physical mechanism that provides the time variation. If the mechanical motion is not specified in advance, we cannot use this model; instead we must use a nonlinear time-invariant model that includes mechanical as well as electrical variables.

Another example might be helpful. If a nonlinear capacitor is pumped strongly by current at frequency ω_p, and if there are superimposed on this small currents at frequencies $\omega_s + n\omega_p$, where n is a positive or negative integer, we may wish to restrict our attention to the small-signal variables, without the pump. We may do so by using a linear time-varying model, with capacitance C(t) just equal to the incremental capacitance of the nonlinear capacitor, as pumped. Such a technique is universally used in descriptions of frequency converters, because the linear time-varying equations are easier to solve than the original nonlinear ones. This "linearization" process is described in more detail in Sec. 8.2.

A nonlinear capacitor is not lossless. By "lossless" we mean that the time-average power input vanishes:

$$\Sigma_a P_a = 0 \tag{4.31}$$

The time-varying capacitor does not obey Eq. 4.31, so it cannot be said to be lossless. Net power can be generated or destroyed by this model. Of course, the actual physical device may be lossless, in which case this energy is accounted for by loss or gain of energy by the "prime mover" or "pump," but this fact is not predicted by the time-varying model.

Simply because the time-varying capacitor is not lossless, we do not wish to imply that it is lossy, that is, having dissipative loss. One connotation often given to the word "lossless" is freedom from dissipation, with its noise, entropy increase, etc. This is not exactly what we mean here.

It is stated in Chapter 1 that losslessness is neither necessary nor sufficient for the Manley-Rowe formulas to hold. The example of Appendix A shows that it is not sufficient; the example here shows that it is not necessary: the time-varying capacitor, although not lossless, obeys the frequency-power formulas.

Now we show that the nonlinear, time-varying capacitor obeys the formulas of Types I and IV (although, of course, only those formulas corresponding to an independent frequency ω_a, on which the explicit time variation does not depend). The nonlinear, time-varying constitutive relation is

$$e = e(q, t) \tag{4.32}$$

and the reasoning in Sec. 3.5 leads to frequency-power formulas of Type IV, in the form of Eq. 4.16, with the Q_α given by Eq. 4.17. To show that the device obeys the Manley-Rowe formulas, we define the time-varying energy function

$$W_c(q, t) = \int_0^q e(q, t)\, dq \tag{4.33}$$

This energy function leads to formulas of the form of Eq. 4.14, with the P_α given, as expected, by Eq. 4.15.

Similar results hold for the time-varying inductor.

4.5 Mechanical Elements

Frequency-power formulas of Types I and IV hold for nonlinear, time-varying mechanical energy-storage elements. We do not investigate Type IV, since reactive power is not often used with mechanical elements.

A nonlinear, possibly time-varying spring has a constitutive relation

$$f = f(x, t) \tag{4.34}$$

where f is the force exerted on the spring and x is its displacement. Although we use one-dimensional motion in the example, the notions are readily extended to more than one dimension, and to rotational systems. A simple example of a nonlinear (but time-invariant) constitutive relation like Eq. 4.34 is that between torque and angular displacement of a pendulum,

$$\tau = mgr \sin\theta \tag{4.35}$$

The potential energy

$$W_p(x, t) = \int f(x, t)\, dx \tag{4.36}$$

is an energy-state function, and its use leads to frequency-power formulas in the form of Eq. 4.14, with P_α interpreted as the mechanical power input at frequency ω_α:

$$P_\alpha = \tfrac{1}{2}\mathrm{Re}\, f_\alpha^* v_\alpha \tag{4.37}$$

where v is the velocity, the time derivative of the displacement.

In a similar way, one can prove the Manley-Rowe formulas for a nonlinear, time-varying point mass with a constitutive relation

$$v = v(p, t) \tag{4.38}$$

where p is the momentum of the particle, or the time integral of the force. The use of the kinetic-energy function

$$W_k(p, t) = \int v(p, t)\, dp \tag{4.39}$$

leads to frequency-power formulas of the form of Eq. 4.14, with the P_a given by Eq. 4.37.

It might be argued that in reality the constitutive relation of Eq. 4.38 for point masses is linear, in the form

$$p = mv \tag{4.40}$$

where m is the mass, and so the Manley-Rowe formulas hold in a trivial way, with all $P_a = 0$. This is true for nonrelativistic particles, but for relativistic speeds Eq. **4.40** is modified to be nonlinear:

$$v = \frac{pc}{(p^2 + m^2c^2)^{1/2}} \tag{4.41}$$

4.6 Rotating Machines

Some rotating machines obey the Manley-Rowe formulas. These machines are those without commutators; that is, they are suitable for use as induction, synchronous, or reluctance machines.

As a model for these machines, we consider a set of coils — some of them fixed (on the stator) and others movable (on the rotor). We neglect losses like those due to winding resistance, eddy currents, mechanical friction, and hysteresis. The rotor and stator shape need not be specified, so salient-pole machines are included. Neither the number of windings nor their configuration is specified, so we can have any number of phases and any number of poles. The magnetic material can even be nonlinear (although it must be free from hysteresis and dispersion). This model is a bit more general than the "generalized machine" [116, Ch. 3], which has been proposed as a model for practical magnetic-field rotating devices without commutators.

The coil currents i_k (k is an index) and shaft position can be considered as independent variables to define the magnetic coenergy [116, Ch. 3] by the line integral

$$W_m'(i_\ell, \phi) = \Sigma_k \int \lambda_k\, di_k \tag{4.42}$$

so that

$$\lambda_k = \frac{\partial W_m'(i_\ell, \phi)}{\partial i_k} \tag{4.43}$$

is the flux linkage in the k-th coil, and

$$\tau = -\frac{\partial W_m'(i_\ell, \phi)}{\partial \phi} \tag{4.44}$$

is the torque applied to the shaft [116, Ch. 3].

The shaft speed $\omega_m = \frac{d\phi}{dt}$ can have an average value, so if ϕ is written in the form of Eq. 3.17, v_{i0} need not vanish. There can be d-c power input at the shaft, although not at any of the coil terminals. It is primarily with this application in mind that provision is made for d-c power terms in the Manley-Rowe formulas.

The magnetic coenergy is a suitable state function for deriving frequency-power formulas of Type I; they are of the form

$$\frac{P_0}{\omega} \frac{\partial \omega}{\partial \omega_a} + \Sigma_\alpha \frac{P_\alpha}{\omega_\alpha} \frac{\partial \omega_\alpha}{\partial \omega_a} = 0 \tag{4.45}$$

where ω is the average value of ω_m, P_0 is the average torque times ω, and each P_α has both electrical and mechanical contributions:

$$P_\alpha = \tfrac{1}{2}\text{Re}\left[\tau_\alpha^*(\omega_m)_\alpha + \Sigma_k(e_k)_\alpha(i_k)_\alpha^*\right] \tag{4.46}$$

It is evident from Eq. 4.45 that no error would be made if the d-c power input at the shaft were considered as being at frequency ω. This is true in general for rotating shaft d-c power, as proved in Appendix D.

These rotating machines also obey frequency-power formulas of Type IV, provided the frequency ω_a does not appear in the actual time dependence of the shaft angle ϕ. The overall derivative matrix of the form of Eq. 3.66 is not positive definite, because the diagonal term relating the mechanical variables,

$$\frac{\partial \tau(i_\ell, \phi)}{\partial \phi} \tag{4.47}$$

is not necessarily positive. The electric submatrix, however, with i-k entry

$$\frac{\partial \lambda_i(i_\ell, \phi)}{\partial i_k} \tag{4.48}$$

is ordinarily positive definite, so the rotating machines satisfy the "weak positive-definiteness" condition discussed in Sec. 3.5, and therefore obey formulas of Type IV for each ω_a that does not appear in the actual time dependence of $\phi(t)$. The formulas are written in the form of Eq. 4.22, with Q_α the overall reactive power input at frequency ω_α, to all coils,

$$Q_\alpha = \tfrac{1}{2}\,\text{Im}\,\Sigma_k(e_k)_\alpha(i_k)_\alpha^* \tag{4.49}$$

These formulas, of Type IV, can be used [86, 85] to show that the induction generator is stable without some source of reactive power.

Rotating machines with commutators do not ordinarily obey these frequency-power formulas, because the commutators are switches, or time-varying resistors. However, in some cases where it is possible to treat the machine as if these switches were "outside", the formulas hold. This is done, for example, in Sec. 7.3.

4.7 Generalized Conservative Energy Conversion Device

A logical generalization of the rotating machine is the generalized conservative energy conversion device [116, Sec. 1.4], which is a model that describes many physical systems with either magnetic field or electric field energy conversion, or both. This model applies to such physical devices as rotating machines, phonograph cartridges, microphones (except carbon microphones), loudspeakers, solenoids, meter movements, etc. The general model is described by a Lagrangian [116, Sec. 1.4], and this state function can be used to obtain the Manley-Rowe formulas. The expressions for P_α must include mechanical as well as electrical variables.

4.8 Traditors

There are not many idealized nonlinear circuit elements. Duinker [20] has found a set of elements, "traditors," that can be described by a Lagrangian, but do not store energy, so cannot be described by a Hamiltonian. He defines several classes of traditors, but the most important seem to be the "third-degree" traditors, the simplest to be actually nonlinear. He shows how higher-degree and lower-degree traditors can be synthesized from third-degree traditors. We show here that traditors of the third degree (and hence of any degree) obey frequency-power formulas of Type I, a result also obtained by Duinker [18].

Six distinct forms of third-degree traditors are discussed by Duinker [20], but any can be constructed from any other one and gyrators. Thus, it is necessary to discuss only one form here — we take "form I."

A third-degree traditor of form I is a three-port device with equations of motion

$$e_1 = -A q_2 i_3 \tag{4.50}$$

$$e_2 = -A q_1 i_3 \tag{4.51}$$

$$e_3 = A q_1 i_2 + A q_2 i_1 \tag{4.52}$$

where e_k, i_k, and q_k are respectively the voltage, current, and charge at the k-th port. The constant A is the only device parameter. This device has an instantaneous power input of zero, just like the gyrator and ideal transformer. However, unlike the ideal transformer and the gyrator, it is nonlinear, so the power inputs at the various frequencies need not vanish.

The device has a Lagrangian

$$L = A q_1 q_2 i_3 \tag{4.53}$$

which is a function of q_1, q_2, and i_3. Its partial derivatives are

$$\frac{\partial L}{\partial q_1} = -e_1 \tag{4.54}$$

$$\frac{\partial L}{\partial q_2} = -e_2 \tag{4.55}$$

and

$$\frac{\partial L}{\partial i_3} = A q_1 q_2 = \int e_3 dt \tag{4.56}$$

and so the use of $-L$ as the energy state function gives frequency-power formulas in the form of Eq. 4.14, with

$$P_\alpha = \tfrac{1}{2}\text{Re}\, j\omega_\alpha\left[\left(\frac{\partial(-L)}{\partial q_1}\right)_\alpha^* (q_1)_\alpha + \left(\frac{\partial(-L)}{\partial q_2}\right)_\alpha^* (q_2)_\alpha + \left(\frac{\partial(-L)}{\partial i_3}\right)_\alpha^* (i_3)_\alpha\right]$$

$$= \tfrac{1}{2}\text{Re}\left[(e_1)_\alpha^*(i_1)_\alpha + (e_2)_\alpha^*(i_2)_\alpha + (e_3)_\alpha^*(i_3)_\alpha\right] \tag{4.57}$$

simply the power input to all ports at frequency ω_α. Although traditors connected together will pass d-c power, this cannot happen if all the variables are written in the forms of Eqs. 3.15, 3.16, and 3.17. We need not, therefore, bother with the d-c power terms of the Manley-Rowe formulas.

Traditors do not, generally, obey frequency-power formulas of Type IV, because they are not locally passive.

4.9 Degenerate Elements

In this section we discuss some elements that obey the frequency-power formulas in a trivial way. First we discuss the ideal transformer, and then the gyrator. Next we give a generalization, and then apply it to a gear system, such as a differential. The results formally apply to the maser, as we point out at the end, but they do not give much physical insight.

Ideal Transformer. By the term, "ideal transformer," we mean a two-port device with equations

$$e_2(t) = Ne_1(t) \tag{4.58}$$

and

$$i_1(t) = -Ni_2(t) \tag{4.59}$$

where the constant N is called the "turns ratio." Except for trivial cases ($N = 0$, ± 1, or ∞), the model is approximated physically by a pair of coils on a common magnetic core. With perfect coupling between the coils, the flux linking one also links the other, so the two flux linkages (and therefore the voltages) differ only by a constant (the turns ratio), so Eq. 4.58 holds. The constitutive relation of the core material gives a relation between $(i_1 + Ni_2)$ and the flux linkage λ_2, and without hysteresis or other losses this might be approximated by a curve like Fig. 4.2. If the excitation is such that $|\lambda_2|$

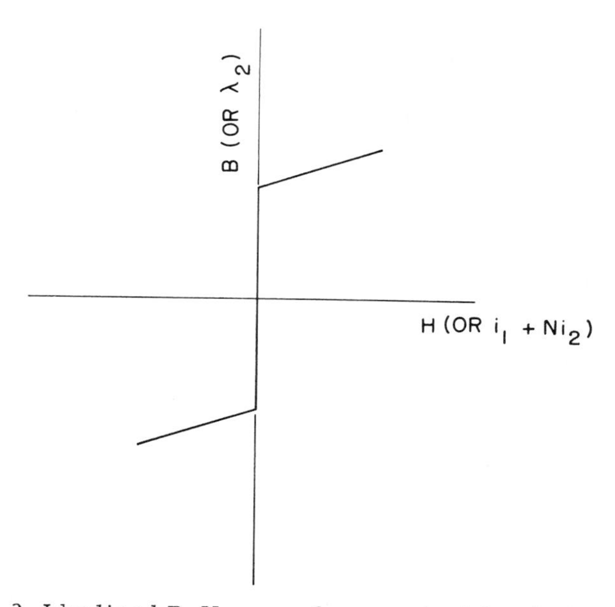

Fig. 4.2 Idealized B-H curve for a material suitable for a transformer. For values of λ_2 less than the critical amount, the magnetic field is vanishingly small. To keep the flux below this critical value, we must avoid having a non-zero average value of voltage; hence the physical transformer (using this material) cannot pass d-c power

is always less than the critical value shown in Fig. 4.2, then $i_1 + Ni_2$ vanishes, so Eq. 4.59 holds. If, however, $|\lambda_2|$ gets too large, then Eq. 4.59 does not hold, and the device is not an ideal transformer, because it has saturated. The physical transformer cannot pass d-c power, but the ideal transformer (which is only a model) can.

The physical transformer is merely a pair of nonlinearly-coupled coils, and from the results of Sec. 4.3 we know it obeys frequency-power formulas of Types I and IV. We now show that the ideal transformer model also satisfies these formulas, but in a trivial way.

The vector power into the ideal transformer at frequency ω_α is

$$P_\alpha + jQ_\alpha = \frac{1}{2}\left[(e_1)_\alpha (i_1)_\alpha^* + (e_2)_\alpha (i_2)_\alpha^*\right] \qquad (4.60)$$

and because Eqs. 4.58 and 4.59 hold in frequency-by-frequency form, Eq. 4.60 must vanish. Thus all Q_a vanish, so any sum of them vanishes, and so in particular frequency-power formulas of Type IV, with the inequality sign in either direction, hold.

By a similar argument, one can easily see that the d-c power sum of Eq. 3.36 vanishes, and since all P_a vanish, Eqs. 3.36, the Manley-Rowe formulas, hold.

Gyrator. The gyrator [104, 94] is a nonreciprocal circuit model with two ports, and equations of motion

$$e_2(t) = Ri_1(t) \tag{4.61}$$

and

$$e_1(t) = -Ri_2(t) \tag{4.62}$$

Like the ideal transformer, the gyrator does not store or dissipate any energy, but unlike the transformer it can accept arbitrary amounts of reactive power, of either sign. It obeys the Manley-Rowe formulas but does not obey frequency-power formulas of Type IV.

The vector power at frequency ω_a is

$$\begin{aligned} P_a + jQ_a &= \tfrac{1}{2}\left[(e_1)_a(i_1)_a^* + (e_2)_a(i_2)_a^*\right] \\ &= \tfrac{1}{2}R\left[(i_1)_a(i_2)_a^* - (i_2)_a(i_1)_a^*\right] \end{aligned} \tag{4.63}$$

and therefore is pure imaginary. Thus all P_a vanish, and because one can show easily that the d-c part of Eq. 3.36 vanishes, the frequency-power formulas of Type I, Eq. 3.36, hold, with the P_a interpreted as the net power input to the gyrator at frequency ω_a.

A Generalization. These results can be generalized somewhat. Suppose a device cannot store or dissipate energy so the instantaneous power input

$$P(t) = \Sigma_i f_i(t) v_i(t) = 0 \tag{4.64}$$

where the f_i variables do not depend at all on the v_i variables; and suppose that the v_i variables are not all independent, so that they can be written in terms of a smaller number of independent variables v'_k, in a linear time-invariant way

$$v_i(t) = \Sigma_k C_{ik} v'_k(t) \tag{4.65}$$

For simplicity let us use matrix notation; let $\underset{\sim}{v}(t)$ be the column matrix of $v_i(t)$; let $\underset{\sim}{v}'(t)$ be the column matrix of $v'_k(t)$; let $\underset{\sim}{f}(t)$ be the column matrix of the $f_k(t)$; and let $\underset{\sim}{C}$ be the rectangular matrix with entries C_{ik}. Then,

$$\underset{\sim}{v} = \underset{\sim}{C}\underset{\sim}{v}' \tag{4.66}$$

and

$$0 = \underset{\sim}{f}_t \underset{\sim}{v}$$

$$= \underset{\sim}{f}_t \underset{\sim}{C} \underset{\sim}{v}' \tag{4.67}$$

where the t subscript indicates the transpose. Since the v'_k variables are independent, we must have

$$\underset{\sim}{f}_t \underset{\sim}{C} = 0 \tag{4.68}$$

If we express all variables in the form of Eqs. 3.15 and 3.16, we may derive from Eqs. 4.66 and 4.68 the equations relating the average values of the variables,

$$\underset{\sim}{v}_0 = \underset{\sim}{C} \underset{\sim}{v}_0' \tag{4.69}$$

$$\underset{\sim}{f}_{0t} \underset{\sim}{C} = 0 \tag{4.70}$$

and the equations relating each set of Fourier coefficients,

$$\underset{\sim}{v}_\alpha = \underset{\sim}{C} \underset{\sim}{v}_\alpha' \tag{4.71}$$

and

$$\underset{\sim}{f}_\alpha{}^\dagger \underset{\sim}{C} = 0 \tag{4.72}$$

where the dagger $\dagger$ indicates the complex conjugate transpose of a matrix.

The d-c terms in the Manley-Rowe formulas of Eq. 3.36 are written, in our notation,

$$\begin{aligned} \Sigma_i \frac{P_{i0}}{v_{i0}} \frac{\partial v_{i0}}{\partial \omega_a} &= \underset{\sim}{f}_{0t} \frac{\partial \underset{\sim}{v}_0}{\partial \omega_a} \\ &= \underset{\sim}{f}_{0t} \underset{\sim}{C} \frac{\partial \underset{\sim}{v}_0'}{\partial \omega_a} \\ &= 0 \end{aligned} \tag{4.73}$$

and the vector power at each frequency ω_α is

$$\begin{aligned} P_\alpha + jQ_\alpha &= \tfrac{1}{2}(\underset{\sim}{f}_\alpha{}^\dagger \underset{\sim}{v}_\alpha) \\ &= \tfrac{1}{2}\underset{\sim}{f}_\alpha{}^\dagger \underset{\sim}{C} \underset{\sim}{v}_\alpha' \\ &= 0 \end{aligned} \tag{4.74}$$

and so the Manley-Rowe formulas, Eq. 3.36, hold in a trivial way, and the frequency-power formulas of Type IV hold, also in a trivial way, provided Q_α can be interpreted as reactive power.

This generalization clearly applies to both the ideal transformer and the gyrator. In the former case the Q_α can be interpreted as reactive powers, and in the latter case they cannot. We now show how this generalization applies to gear systems, and then, formally, to masers.

Gear Systems. If we neglect energy storage and friction in gear systems, the preceeding generalization applies to them. A typical gear system of interest is the differential; we use it as an example. The shaft speeds are identified with the v_i variables, and the shaft torques τ with the f_i variables. The speed constraint for the differential is in the form

$$\omega_1 = r(\omega_3 - \omega_2) \tag{4.75}$$

Letting ω_2 and ω_3 be independent variables, Eq. 4.66 becomes

$$\begin{bmatrix} \omega_1 \\ \omega_2 \\ \omega_3 \end{bmatrix} = \begin{bmatrix} -r & r \\ 1 & 0 \\ 0 & 1 \end{bmatrix} \times \begin{bmatrix} \omega_2 \\ \omega_3 \end{bmatrix} \tag{4.76}$$

and the condition on torques, Eq. 4.68, is

$$[\tau_1 \ \tau_2 \ \tau_3] \times \begin{bmatrix} -r & r \\ 1 & 0 \\ 0 & 1 \end{bmatrix} = [0 \ \ 0] \tag{4.77}$$

If only d-c power is present, then the Manley-Rowe formulas reduce to the form of Eq. 1.16, which is well known.

It is interesting to note that in Eq. 1.16 all the power flow is d-c, yet the form of the equations is what one would expect from a-c power terms. This suggests that one may actually consider d-c power flow terms at a rotating shaft to be power at frequency equal to the shaft speed. This is generally possible, as we show in Appendix D.

Maser. The solid-state maser formally is described by the generalization of Sec. 4.9, although this does not provide much physica insight. The results are obtained more easily by the method of Weiss[115].

In practice, the "idler" power given off by a maser is absorbed by losses in the crystal, as is a majority of the pump power. If, hov ever, these losses are considered as being "outside" the system, th frequency-power formulas of Type I apply.

Let us identify the energy level separations E_i with the v_i vari ables, and the number of photons per second entering the system at frequency $\nu_i = E_i/h$, where h is Plank's constant, with the f_i variables. The energy level separations are not all independent, for certain of them add together to equal certain other ones. In the simple three-level maser, the two smaller energy level separations added together just equal the large one. An independent set of E_i may be extracted, or, what is equivalent, an independent set of ν_i. There is a constraint among the number of photons entering the system at each frequency; in the three-level maser the number at each frequency must be equal (in magnitude). The formal application of the generalization of Sec. 4.9 leads to frequency-power formulas similar to Eq. 1.13, which holds for the three-level maser. Unfortunately, this treatment does not provide much physical insight; it is merely by mathematical analogy.

4.10 Network Extension

Suppose we have a network of elements like nonlinear, time-varying capacitors and inductors, rotating machines, ideal transformers, and other devices that obey the Manley-Rowe formulas, and suppose this network has a number of terminal pairs, through which power flows to the network. If the entire network is to be used for frequency conversion, the quantities of engineering interest are the powers at each frequency at the ports of the network. Even though each element in the network obeys the Manley-Rowe formulas, we do not as yet know that the network as a whole does, in such a way that the powers P_α can be evaluated at the ports. This is what is proved in this section.

A similar extension of frequency-power formulas of Type IV is possible if the ambiguity of sign of the reactive power can be accounted for. The extension holds if the devices are either all capacitive or all inductive, but it is not valid if the network contains both capacitive and inductive elements.

These extensions are intuitively reasonable, and are proven here by general topological arguments. Let the terminal pairs of the individual elements in the network be indexed by k. Then the overall vector power into all the elements at frequency ω_α is

$$P_\alpha + jQ_\alpha = \tfrac{1}{2}\Sigma_k (e_k)_\alpha (i_k)_\alpha^* \tag{4.78}$$

If each element obeys frequency-power formulas of Type I, then by adding together the formulas one arrives at formulas of the same form with the P_α of Eq. 4.78, and alternatively, if each element satisfies formulas of Type IV (all with the same direction of inequality), then by adding together formulas one arrives at formulas of the same form with the Q_α of Eq. 4.78. What we show here is that the $P_\alpha + jQ_\alpha$ are identical to the vector powers entering the network at its ports. Therefore, the network obeys the appropriate type of frequency-power formulas, with the powers (P_α or Q_α) evaluated at the ports of the network.

To show that $P_\alpha + jQ_\alpha$ is the vector power into the network at frequency ω_α, we use a topological argument [33, Ch. 1]. We represent multiport devices within the network by a scheme proposed by Guillemin [34], and for topological purposes we consider each port of the network to be a separate branch, with voltage e_ℓ and current $-i_\ell$, where ℓ is an index over the ports of the network. The branches of the network consist, then, of the elements and the terminal pairs.

We call the column matrix of branch currents $\underset{\sim}{i}$, and the column matrix of branch voltages $\underset{\sim}{e}$. Then one of Kirchhoff's laws is of the form

$$\underset{\sim}{i} = \underset{\sim}{T}_t \underset{\sim}{i}' \tag{4.79}$$

where, picking any tree [33, Ch. 1], we denote the column matrix of link currents by $\underset{\sim}{i}'$. The matrix $\underset{\sim}{T}$ is the tie-set schedule [33, Ch. 1]. The other one of Kirchhoff's laws is

$$\underset{\sim}{T}\underset{\sim}{e} = 0 \tag{4.80}$$

Taking Fourier coefficients of Eqs. 4.79 and 4.80, we find

$$\underset{\sim}{i}_\alpha^\dagger = \underset{\sim}{i}_\alpha'^\dagger \underset{\sim}{T} \tag{4.81}$$

and

$$\underset{\sim}{T}\underset{\sim}{e}_\alpha = 0 \tag{4.82}$$

We wish to show that

$$P_\alpha + jQ_\alpha = \tfrac{1}{2}\Sigma_\ell \left[(e_\ell)_\alpha (i_\ell)_\alpha^*\right] \tag{4.83}$$

This is clear, since

$$\begin{aligned} P_\alpha + jQ_\alpha - \tfrac{1}{2}\Sigma_\ell (e_\ell)_\alpha (i_\ell)_\alpha^* &= \tfrac{1}{2}\Sigma_k (e_k)_\alpha (i_k)_\alpha^* - \tfrac{1}{2}\Sigma_\ell (e_\ell)_\alpha (i_\ell)_\alpha^* \\ &= \tfrac{1}{2}\underset{\sim}{i}_\alpha^\dagger \underset{\sim}{e}_\alpha \\ &= \tfrac{1}{2}\underset{\sim}{i}_\alpha'^\dagger \underset{\sim}{T}\underset{\sim}{e}_\alpha \\ &= 0 \end{aligned} \tag{4.84}$$

Similar reasoning applies to all-mechanical and electric-mechanical networks.

4.11 Linear Time-Invariant Resistors

Linear time-invariant resistors are not reactive devices, but they are included here because they obey frequency-power formulas of Type IV. No reactive power flows in at any frequency, so all Q_α vanish. The formulas then are satisfied in a trivial way, and linear resistors can be included in networks that are intended to obey formulas of Type IV.

Chapter 5

LUMPED DISSIPATIVE SYSTEMS

In this chapter we shall discuss some systems that satisfy frequency-power formulas of Types II and III. Generally speaking, such systems are dissipative, although the degenerate cases discussed in Secs. 5.3 and 5.5 are not.

In Sec. 5.1 nonlinear resistors are treated, and in Sec. 5.2 time-varying resistors. Some degenerate devices appear in Sec. 5.3, and in Sec. 5.4 it is shown that a network, each element of which obeys frequency-power formulas of one type, itself obeys formulas of the same type. In Sec. 5.5 it is pointed out that linear lossless devices obey formulas of Type III.

5.1 Nonlinear Resistors

A nonlinear resistor is characterized by a dissipation function, the content [68]. The voltage e is some function of the current i,

$$e = e(i) \tag{5.1}$$

and so the content, defined by

$$G(i) = \int e(i)\, di \tag{5.2}$$

is a function of the current. In addition, one can define another state function, the co-content [68], by the Legendre transformation

$$G'(e) = ie - G(i) \tag{5.3}$$

A typical volt-ampere curve for a resistor is shown in Fig. 5.1; the state functions correspond to the areas indicated.

In demonstrating that the nonlinear resistor obeys frequency-power formulas of the second type, we may use either the content or the co-content. Let us use the content. Then the conditions set up in Sec. 3.3 are satisfied, and we find

$$\begin{aligned} 0 &= \Sigma_\alpha \frac{j}{4} \frac{\partial \omega_\alpha}{\partial \omega_a} (e_\alpha{}^* i_\alpha - i_\alpha{}^* e_\alpha) \\ &= \Sigma_\alpha \frac{\partial \omega_\alpha}{\partial \omega_a} Q_\alpha \end{aligned} \tag{5.4}$$

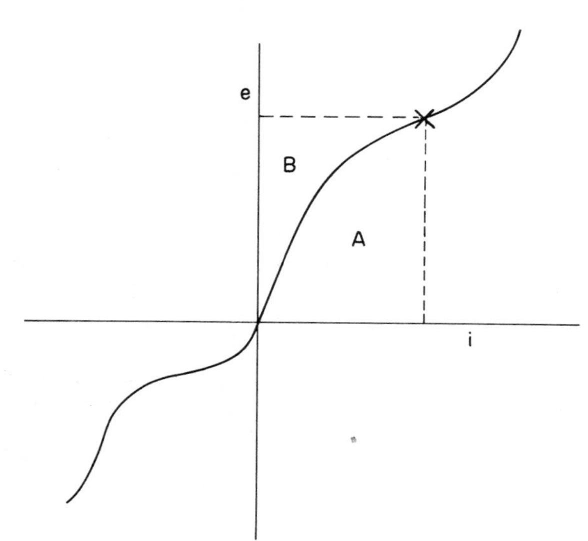

Fig. 5.1. A typical voltage-current characteristic of a nonlinear resistor. At the operating point indicated, the content and co-content are represented by the areas A and B respectively

where

$$Q_\alpha = \tfrac{1}{2}\mathrm{Im}\, e_\alpha i_\alpha^* \tag{5.5}$$

In addition, we know from Sec. 3.4 that because of the functional relation of Eq. 5.1, the device satisfies frequency-power formulas of the third type, in any of the forms derived in Sec. 3.4, such as Pantell's form

$$h_a = \Sigma_\alpha \left(\frac{\partial \omega_\alpha}{\partial \omega_a}\right)^2 P_\alpha \geq 0 \tag{5.6}$$

where

$$P_\alpha = \tfrac{1}{2}\mathrm{Re}\, e_\alpha i_\alpha^* \tag{5.7}$$

provided that the incremental resistance

$$R_i = \frac{de}{di} \tag{5.8}$$

is nowhere negative.

A nonlinear resistor of great theoretical importance is the ideal diode, whose volt-ampere curve is shown in Fig. 5.2. This device is lossless (because at all times either the voltage, the current, or

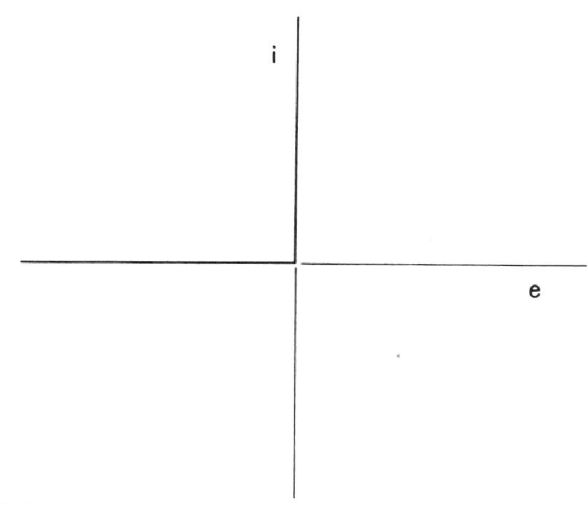

Fig. 5.2. Volt-ampere curve of an ideal diode. Although the device is lossless, its operation is described by a relation between voltage and current, so it is a nonlinear resistor

both vanish), but it is properly classed as a nonlinear resistor, because its operation is defined in the voltage-current plane. It satisfies frequency-power formulas of both Types II and III.

As far as frequency conversion is concerned, the most important physical devices that satisfy these formulas are point-contact diodes, or crystal diodes [106, Ch. 5; 108; 66; 107], which are in wide use in communications systems as mixers, modulators, etc. Some other physical devices that are well approximated as nonlinear resistors are rectifiers, varistors, field-emission diodes [26], and special-purpose semiconductor diodes, including very narrow base diodes [84], avalanche (Zener) diodes [63], and current limiters [111]. All these obey formulas of both Types II and III.

A few nonlinear resistors have a "negative-resistance" region, and therefore cannot be expected to obey formulas of Type III, although they do obey formulas of Type II. Typical are the tunnel diode [22, 97] (if the barrier capacitance is neglected), the p-n-p-n switching diode [69, 95, 96], etc. [81].

Some nonreciprocal two-port nonlinear resistors obey formulas of Type III, but not Type II, since the dissipation fuction is not defined. Included are the Hall effect gyrator [32, 61], and other chunks of ohmic material, with two or more ports, immersed in a magnetic field.

Nonlinear friction devices, including dashpots and ratchets, also obey the formulas, as one can see by analogy.

5.2 Time-Varying Resistors

Time variation of a resistance can arise in many ways. Probably the simplest time-variable resistor is one that is varied mechanically — a potentiometer, and, as a limiting case, a switch. Since, however, energy is not required to vary a resistor (contrary to the case of a capacitor or an inductor), many physical effects can be used.

Resistors that are varied mechanically include potentiometers, rheostats, switches, and strain gages. Carbon microphones and piezoresistive microphones [9] are nothing more than resistors whose instantaneous values depends on the sound pressure. Thermistors (and in fact, all resistors) have a resistance that depends on temperature, and some photoconductors can be thought of as linear resistors whose resistance depends on the light intensity. The resistance matrix of a chunk of ohmic material immersed in a magnetic field depends on the strength and orientation of the field. As a last example, the magnetic field dependence of superconductivity has been used to make time-varying resistors (cryotrons) [8], and these have been considered for possible frequency conversion applications [118].

Just as it is not necessary in Sec. 4.4 to think of a time-varying reactance in terms of the physical effect that produces the variation, so here it is not necessary to determine whether the time variation is caused by any particular physical effect, or by heavy nonlinear pumping (see Sec. 8.2).

A nonlinear time-varying resistor has a voltage that is a function of current and also explicitly of time

$$e = e(i, t) \tag{5.9}$$

and so one can define a time-varying dissipation function

$$G(i, t) = \int e(i, t)\, di \tag{5.10}$$

and the proof of the second type of frequency-power formulas, in Sec. 3.3, predicts that

$$0 = \Sigma_\alpha \frac{\partial \omega_\alpha}{\partial \omega_a} Q_\alpha \tag{5.11}$$

where

$$Q_\alpha = \tfrac{1}{2} \operatorname{Im} e_\alpha i_\alpha^* \tag{5.12}$$

for each independent frequency ω_a that is not used to help specify the explicit time variation of G.

Similarly, frequency-power formulas of the third type hold in any of the forms discussed in Sec. 3.4, including the form

$$h_a = \Sigma_\alpha \left(\frac{\partial \omega_\alpha}{\partial \omega_a} \right)^2 P_\alpha \geq 0 \tag{5.13}$$

where

$$P_\alpha = \tfrac{1}{2}\mathrm{Re}\; e_\alpha i_\alpha^* \tag{5.14}$$

provided the excitation is such that at no time is the time-varying incremental resistance

$$R_i(t) = \frac{\partial e(i, t)}{\partial i} \tag{5.15}$$

negative. Equation 5.13 holds for each independent frequency ω_α that is not used to help specify the explicit time dependence of $e(i, t)$.

By analogy, time-varying dashpots and ratchets and periodically-applied mechanical clamps obey the formulas of Types II and III, just as their electrical counterparts do.

It is interesting to note that a linear time-invariant resistor will accept no reactive power at any frequency. A nonlinear resistor can accept net reactive power, and is capable of accepting it at some frequencies and releasing it at others, but there is a sum of reactive powers that vanishes:

$$\Sigma_\alpha \omega_\alpha Q_\alpha = 0 \tag{5.16}$$

A time-varying resistor, on the other hand, can generate or accept net reactive power, and the sum of Eq. 5.16 is not conserved. The mechanism that produces the time variation of the resistance can be thought of as a source of reactive power.

This situation is similar to that of reactances. A linear time-invariant reactance will not accept real power at any frequency; a nonlinear time-invariant reactance will accept real power at some frequencies and release it at others, although the overall real power input must vanish; and a time-varying reactance can generate or accept net amounts of real power. The physical mechanism that varies the reactance can be thought of as a source of real power.

5.3 Degenerate Elements

Generally speaking, the degenerate devices discussed in Sec. 4.9 obey frequency-power formulas of Types II and III in the same trivial way they obey formulas of Types I and IV. The ideal transformer obeys both Types II and III, with all P_α and Q_α zero. The ideal gyrator obeys formulas of Type III (but not Type II) with all P_α zero. The other degenerate elements are treated in a similar way.

As mentioned in Sec. 4.9 (Ideal Transformer), a physical transformer cannot pass d-c power. If it is placed in a circuit with a non-zero average voltage, it will saturate, and no longer approximate an ideal transformer. Nonlinear resistors often have d-c power input, and if a transformer is placed in a circuit with them, some alternate d-c path must be provided if the transformer is to avoid saturation.

It is interesting to note that a time-varying transformer, such as an adjustable autotransformer, obeys frequency-power formulas of Types II and III. This device obeys equations of motion like Eqs. 4.58 and 4.59 with the turns ratio N a function of time:

$$e_2(t) = N(t)e_1(t) \tag{5.17}$$

and

$$i_1(t) = -N(t)i_2(t) \tag{5.18}$$

It is easily verified that this device is lossless and linear, but since it is time-varying the power input at each frequency P_α need not vanish. The device does not obey frequency-power formulas of Types I or IV, but does obey Types II and III. Thus, combinations of variable transformers and potentiometers [87] obey these formulas also.

To show that this device satisfies formulas of Type III, we consider e_2 and i_1 to be functions of e_1 and i_2, given by Eqs. 5.17 and 5.18. The matrix

$$\begin{bmatrix} \dfrac{\partial e_2}{\partial i_2} & \dfrac{\partial e_2}{\partial e_1} \\ \\ \dfrac{\partial i_1}{\partial i_2} & \dfrac{\partial i_1}{\partial e_1} \end{bmatrix} = \begin{bmatrix} 0 & N(t) \\ \\ -N(t) & 0 \end{bmatrix} \tag{5.19}$$

is positive definite, so the proof of Sec. 3.4 indicates that

$$h_a = \Sigma_\alpha \left(\frac{\partial \omega_\alpha}{\partial \omega_a} \right)^2 P_\alpha \geq 0 \tag{5.20}$$

and any of the other forms of frequency-power formulas of Type III hold, where P_α is the power input at frequency ω_α,

$$P_\alpha = \tfrac{1}{2}\mathrm{Re}\left[(e_1)_\alpha (i_1)_\alpha^* + (e_2)_\alpha (i_2)_\alpha^* \right] \tag{5.21}$$

and where ω_a is, as usual, an independent frequency on which the time variation of $N(t)$ does not depend.

To show that this device satisfies formulas of Type II, we consider e_2 and $-i_1$ to be functions of e_1 and i_2, as in Eqs. 5.17 and 5.18. Nothing that

$$\frac{\partial e_2}{\partial e_1} = \frac{\partial(-i_1)}{\partial i_2} = N(t) \tag{5.22}$$

we define the time-varying dissipation function

$$\begin{aligned} G(e_1, i_2, t) &= \int e_2 di_2 - \int i_1 de_1 \\ &= N(t) e_1 i_2 \end{aligned} \tag{5.23}$$

independent of the path of integration. The use of this state function yields frequency-power formulas of Type II,

$$\Sigma_\alpha \frac{\partial \omega_\alpha}{\partial \omega_a} Q_\alpha = 0 \tag{5.24}$$

where Q_α is the reactive power input at frequency ω_α,

$$Q_\alpha = \tfrac{1}{2}\mathrm{Im}\left[(e_1)_\alpha (i_1)_\alpha^* + (e_2)_\alpha (i_2)_\alpha^*\right] \tag{5.25}$$

In a similar way one can show that a time-varying gyrator, that is, a device that satisfies equations of motion of the form of Eqs. 4.61 and 4.62 with R a function of time, obeys frequency-power formulas of Type III but not Type II.

5.4 Network Extension

If we have a network of elements, each of which individually satisfies frequency-power formulas of Type II (or III), it is of interest to ask whether the entire network does, in such a way that the powers P_α and Q_α can be evaluated at the ports of the network. The answer is that it does, and the proof of this is identical to the proof given in Sec. 4.10, for networks of elements that obey formulas of either Type I or IV.

5.5 Linear Time-Invariant Lossless Elements

Generally speaking, the devices that obey frequency-power formulas of Types II and III are dissipative devices, or limits of dissipative devices. However, because linear time-invariant lossless elements cannot perform frequency conversion, the power input to such devices at each frequency, P_α, is zero. They therefore satisfy frequency-power formulas of Type III in a trivial way.

Chapter 6

DISTRIBUTED SYSTEMS

In this chapter we deal with frequency-power formulas of Type I only, except that in Sec. 6.1 (Type III Formulas) formulas of Type III are derived. Formulas of Types II and IV do not apply because the Q_α quantities appearing in the formulas cannot be interpreted as reactive powers.

Distributed systems are those with important space variation. Energy is stored and dissipated throughout the space, but power enters and leaves only at the boundary, through the ports. Notice the similarity to a network of lumped elements, in which energy is stored and/or dissipated throughout the system, but power enters only at the ports. In Secs. 4.10 and 5.4 it is shown that, for the purpose of applying frequency-power formulas of Types I and III, the powers P_α can be evaluated at the ports, rather than at each element in the network. Here we show the same for some distributed systems.

The quantities of engineering interest for frequency conversion in a distributed system are the powers entering and leaving the system at each frequency. However, the variables that describe a distributed system are functions of time and space. The constitutive relations and energy state functions are defined in terms of these variables, with explicit space (and perhaps time) variation. The P_α quantities in the frequency-power formulas are functions of space (but not time); their interpretation is not clear.

Suppose, however, that each P_α is a divergence

$$P_\alpha = - \nabla \cdot \overline{F}_\alpha \tag{6.1}$$

where $\overline{F}_\alpha$ can be expressed in terms of the variables of the system. Then the frequency-power formulas of Type I,

$$\Sigma_\alpha \frac{P_\alpha}{\omega_\alpha} \frac{\partial \omega_\alpha}{\partial \omega_a} = 0 \tag{6.2}$$

or Type III, in any of the forms discussed in Sec. 3.4, such as

$$\Sigma_\alpha \left(\frac{\partial \omega_\alpha}{\partial \omega_a} \right)^2 P_\alpha \geq 0 \tag{6.3}$$

whichever are known to hold, can be integrated over the volume of the system, P_a being the only function of space. The result is (by use of the divergence theorem) formulas of Type I

$$\Sigma_a \frac{\iint \overline{F}_a \cdot \overline{n}\, dS}{\omega_a} \frac{\partial \omega_a}{\partial \omega_a} = 0 \tag{6.4}$$

or similar formulas of Type III, such as

$$\Sigma_a \left(\frac{\partial \omega_a}{\partial \omega_a} \right)^2 \iint \overline{F}_a \cdot \overline{n}\, dS \geq 0 \tag{6.5}$$

In Eqs. 6.4 and 6.5, the surface integral is taken over the entire boundary of the system, and $\overline{n}$ is a unit vector pointing inward, perpendicular to the surface. It is our intent that each $\overline{F}_a$ be interpreted as the contribution to time-average value of the power-flow vector $\overline{F}$ caused by frequency ω_a; thus the formulas of Eqs. 6.4 and 6.5 relate quantities of engineering interest, the powers at various frequencies, at the ports of the system.

We conclude, then, that when the P_a are in the form of divergences, the frequency-power formulas relate quantities that can be evaluated at the boundary of the system, the ports. Our problem then reduces to expressing the P_a as divergences. For some physical systems this can be done in a simple way, directly from the equations of motion. Typical are the nonlinear electromagnetic medium discussed by Haus[35], treated here in Sec. 6.1; the irrotational acoustic system discussed by Wagner [110], treated here in Sec. 6.2; and the irrotational electron beam (Haus [36] has discussed the case of most importance, the longitudinal beam with large pump and small signals) of Sec. 6.3.

For systems in which it is difficult to express the P_a in the form of divergences a systematic procedure would be very helpful. Such a procedure is available for systems that obey Hamilton's principle [30, Ch. 11; 70, Part I, Ch. 3, Sec. 3.4], but it is quite complicated. This is not a drawback, though, because once this systematic procedure has indicated the form of the expressions, the results can then be derived directly from the equations of motion. A gyromagnetic medium obeys the full Manley-Rowe formulas, as we show in Sec. 6.4. This fact was originally proved by McWhorter [64] using this systematic procedure (Haus [35] had earlier discussed the case of most practical importance, large pump and small signals), but now that the proper variables are known, a simpler proof can be used.

Hamilton's principle is discussed in Sec. 6.5, along with the systematic procedure leading to the Manley-Rowe formulas. In Sec. 6.6 we discuss isentropic rotational flow of a charged, compressible fluid. This system is so complicated that Hamilton's principle actually simplifies the presentation. Another complicated system, rotational fluid flow with free currents, is discussed in Sec. 6.7; it is an approximation to a magnetohydrodynamic system with infinite conductivity.

One difficulty with this approach is illustrated by these examples. Although the systematic procedure based on Hamilton's principle guarantees that the P_α are in the form of divergences, it does not guarantee that physically unmeasurable potentials or Lagrange multipliers do not appear in the power expressions. Because of this, there is often a real problem in assigning a physical interpretation to the P_α or $\overline{F}_\alpha$.

The form of the expressions for power, when they contain such physically unmeasurable terms, cannot be guessed from a knowledge of the physical power-flow vector. Although any result obtained by use of Hamilton's principle can also be obtained directly from the equations of motion, we would not ordinarily be motivated to define the required physically unmeasurable terms, which are provided automatically by the systematic procedure. The Hamilton's principle method, discussed in Sec. 6.5, is complicated, but its use is justified for systems that cannot be treated directly.

6.1 Electromagnetic Medium

Here we show that the electromagnetic field in a nonlinear, dispersion-free anisotropic, inhomogeneous, reciprocal, stationary electromagnetic medium obeys the Manley-Rowe formulas. This fact was first shown by Haus [35], and our results, in Sec. 6.1 (Manley-Rowe Formulas), are not significantly different. In Sec. 6.1 (Type III Formulas) it is shown that electromagnetic material with nonlinear loss and linear energy storage obeys formulas of Type III.

Equations of Motion. The variables of this system are the electric field $\overline{E}$, the magnetic field $\overline{H}$, the electric displacement vector $\overline{D}$, the magnetic induction vector $\overline{B}$, and the electric current density $\overline{J}$. Each is a function of space and time. We assume that the medium satisfies three constitutive relations: one between $\overline{E}$ and $\overline{D}$; one between $\overline{B}$ and $\overline{H}$; and the third between $\overline{E}$ and $\overline{J}$.

We assume that the magnetic field $\overline{H}$ is a single-valued function of $\overline{B}$ at each point in space. We do not allow $\overline{H}$ to depend on the time or space derivatives of $\overline{B}$, nor on time explicitly, but only on the instantaneous value of $\overline{B}$. Thus, although we do not allow dispersion or hysteresis, the functional dependence may be anisotropic and nonlinear, and may depend on space; for example, the material may be inhomogeneous, or may occupy only a portion of the system. We also do not allow the material to move (that is, it must be stationary), and we require that the dependence be "reciprocal," by which we mean (if Cartesian co-ordinates x, y, and z are used)

$$\frac{\partial H_x}{\partial B_y} = \frac{\partial H_y}{\partial B_x} \tag{6.6}$$

$$\frac{\partial H_y}{\partial B_z} = \frac{\partial H_z}{\partial B_y} \tag{6.7}$$

and

$$\frac{\partial H_z}{\partial B_x} = \frac{\partial H_x}{\partial B_z} \tag{6.8}$$

We have, then, the constitutive relation

$$\overline{H} = \overline{H}(\overline{B}, \overline{r}) \tag{6.9}$$

where we show the explicit dependence on space; because this is reciprocal we may define the "magnetic energy density" by the line integral [42, Sec. 6.10; 80]

$$W_m(\overline{B}, \overline{r}) = \int \overline{H}(\overline{B}, \overline{r}) \cdot d\overline{B} \tag{6.10}$$

In a similar way, we assume that the electric field $\overline{E}$ is a single-valued, nonlinear, dispersion-free, anisotropic, inhomogeneous, reciprocal function of $\overline{D}$

$$\overline{E} = \overline{E}(\overline{D}, \overline{r}) \tag{6.11}$$

so that the line integral

$$W_e(\overline{D}, \overline{r}) = \int \overline{E}(\overline{D}, \overline{r}) \cdot d\overline{D} \tag{6.12}$$

defines the electric energy density.

In portions of the system occupied by free space, the constitutive relations are

$$\overline{H} = \frac{1}{\mu_0} \overline{B} \tag{6.13}$$

and

$$\overline{E} = \frac{1}{\epsilon_0} \overline{D} \tag{6.14}$$

where μ_0 is the permeability of free space and ϵ_0 is the permittivity of free space, and so the energy functions are

$$W_m = \frac{B^2}{2\mu_0} \tag{6.15}$$

and

$$W_e = \frac{D^2}{2\epsilon_0} \tag{6.16}$$

Ohm's law is the constitutive relation between the current density

$\overline{J}$ and the electric field $\overline{E}$. We assume that $\overline{J}$ is a single-valued, nonlinear, dispersion-free, anisotropic, inhomogeneous function of $\overline{E}$ (but one that need not be reciprocal),

$$\overline{J} = \overline{J}(\overline{E}, \overline{r}) \tag{6.17}$$

The power dissipated per unit volume is $\overline{J} \cdot \overline{E}$; if the system is lossless we assume either $\overline{J}$ or $\overline{E}$ vanishes at each point.

The equations of motion are Maxwell's equations

$$\nabla \cdot \overline{B} = 0 \tag{6.18}$$

$$\nabla \cdot \overline{D} = 0 \tag{6.19}$$

$$\nabla \times \overline{H} = \overline{J} + \dot{\overline{D}} \tag{6.20}$$

and

$$\nabla \times \overline{E} = -\dot{\overline{B}} \tag{6.21}$$

These four equations are linear, but the constitutive relations for the system need not be.

From these equations of motion and the definitions of the energy state functions one can derive Poynting's theorem (dot-multiply Eq. 6.20 by $\overline{E}$ and subtract it from Eq. 6.21 dot-multiplied by $\overline{H}$)

$$\frac{\partial}{\partial t}(W_m + W_e) = \overline{H} \cdot \dot{\overline{B}} + \overline{E} \cdot \dot{\overline{D}}$$

$$= -\overline{E} \cdot \overline{J} - \nabla \cdot (\overline{E} \times \overline{H}) \tag{6.22}$$

This is customarily interpreted by identifying $W_e + W_m$ with the energy density, $\overline{E} \cdot \overline{J}$ with the power dissipated per unit volume, and $\overline{E} \times \overline{H}$ with the power-flow vector. Thus, the power dissipated and power going to energy storage must be supplied by the power-flow vector $\overline{E} \times \overline{H}$.

Because Maxwell's equations are linear, we may derive similar theorems involving only Fourier coefficients of the variables. Let us express each variable in the form of Eq. 3.15, where each Fourier coefficient is a complex vector (a set of three complex numbers, or a real vector and an imaginary vector). These complex vectors depend on space, but not on time. Then since Eq. 6.21 holds for the actual vectors $\overline{E}$ and $\overline{B}$, a similar equation holds for each set of Fourier coefficients:

$$\nabla \times \overline{E}_\alpha = -j\omega_\alpha \overline{B}_\alpha \tag{6.23}$$

Similarly, Eq. 6.20 implies equations of the form

$$\nabla \times \overline{H}_\alpha^* = \overline{J}_\alpha^* - j\omega_\alpha \overline{D}_\alpha^* \tag{6.24}$$

If we dot-multiply Eq. 6.24 by $\overline{E}_\alpha$ and subtract it from Eq. 6.23 dot multiplied by $\overline{H}_\alpha^*$, we obtain

$$\nabla \cdot (\overline{E}_\alpha \times \overline{H}_\alpha^*) = -j\omega_\alpha(\overline{B}_\alpha \cdot \overline{H}_\alpha^* - \overline{E}_\alpha \cdot \overline{D}_\alpha^*) - \overline{E}_\alpha \cdot \overline{J}_\alpha^* \tag{6.25}$$

For future reference we note two things. First, if the medium is lossless, then either the current density vanishes and all $\overline{J}_\alpha = 0$, or the electric field vanishes and all $\overline{E}_\alpha = 0$; in either case, the last term in Eq. 6.25 disappears. And secondly, if the energy storage constitutive relations are linear (if $\overline{B}$ is a linear function of $\overline{H}$ and $\overline{D}$ is a linear of $\overline{E}$), then $(\overline{B}_\alpha \cdot \overline{H}_\alpha{}^* - \overline{E}_\alpha \cdot \overline{D}_\alpha{}^*)$ is real, and the first term on the right-hand side of Eq. 6.25 therefore has no real part. The second statement follows from the linear constitutive relations. For example, the relation between $\overline{B}$ and $\overline{H}$ must be of the form

$$B_i = \mu_{ij} H_j \tag{6.26}$$

where B_i and H_j are components of the vectors $\overline{B}$ and $\overline{H}$, and where μ_{ij} is the permeability tensor, and where a sum over the repeated index j is implied. Because $\overline{B}$ is a reciprocal function of $\overline{H}$ the permeability tensor must be reciprocal, so $\mu_{ij} = \mu_{ji}$. Thus

$$\begin{aligned}
\overline{B}_\alpha \cdot \overline{H}_\alpha{}^* &= (B_i)_\alpha (H_i)_\alpha{}^* \\
&= (H_i)_\alpha{}^* (H_j)_\alpha \mu_{ij} \\
&= (H_i)_\alpha{}^* (H_j)_\alpha \mu_{ji} \\
&= (B_j)_\alpha{}^* (H_j)_\alpha \\
&= \overline{B}_\alpha{}^* \cdot \overline{H}_\alpha
\end{aligned} \tag{6.27}$$

and $(\overline{B}_\alpha \cdot \overline{H}_\alpha{}^*)$ is real. In Eq. 6.27 we sum over the repeated indices i and j. Similarly, $(\overline{E}_\alpha \cdot \overline{D}_\alpha{}^*)$ is real.

Manley-Rowe Formulas. Suppose the material is lossless; then $\overline{E}_\alpha \cdot \overline{J}_\alpha{}^*$ vanishes. If we use $W_e + W_m$ as an energy state function for the proof of the Manley-Rowe formulas, we obtain formulas like Eq. 6.2 with the powers P_α given by

$$\begin{aligned}
P_\alpha &= \tfrac{1}{2} \mathrm{Re}\, j\omega_\alpha (\overline{E}_\alpha{}^* \cdot \overline{D}_\alpha + \overline{H}_\alpha{}^* \cdot \overline{B}_\alpha) \\
&= \tfrac{1}{2} \mathrm{Re}\, j\omega_\alpha (\overline{B}_\alpha \cdot \overline{H}_\alpha{}^* - \overline{E}_\alpha \cdot \overline{D}_\alpha{}^*)
\end{aligned} \tag{6.28}$$

and thus by Eq. 6.25

$$P_\alpha = - \nabla \cdot \overline{F}_\alpha \tag{6.29}$$

where

$$\mathbf{F}_\alpha = \tfrac{1}{2} \mathrm{Re}\, \overline{E}_\alpha \times \overline{H}_\alpha{}^* \tag{6.30}$$

The Manley-Rowe formulas can then be written in the form of Eq. 6.4, and the $\overline{F}_\alpha$ quantities can be interpreted as the power-flow vector $\overline{E} \times \overline{H}$ at frequency ω_α. We evaluate the powers at the boundary of the system, as indicated in Eq. 6.4.

Type III Formulas. The system obeys formulas of Type III provided the energy storage is linear. Equation 6.25 becomes

$$\tfrac{1}{2}\mathrm{Re}\,(\overline{E}_\alpha{}^* \cdot \overline{J}_\alpha) = -\nabla \cdot \overline{F}_\alpha \qquad (6.31)$$

where $\overline{F}_\alpha$ is given by Eq. 6.30, and has the interpretation given above.

The constitutive relation of Eq. 6.17 leads to frequency-power formulas of the form of Eq. 6.3 with

$$P_\alpha = \tfrac{1}{2}\mathrm{Re}\,(\overline{E}_\alpha{}^* \cdot \overline{J}_\alpha) \qquad (6.32)$$

Because of Eq. 6.31, these formulas can be integrated to the form of Eq. 6.5, and we can evaluate the powers on the boundary of the system.

6.2 Irrotational Fluid Flow

In this section it is shown that the irrotational flow of a non-viscous fluid obeys the Manley-Rowe formulas. The concept of irrotationality is based on a well-known theorem proved in Appendix F, which states that if a system without viscosity starts flowing so that the curl of the momentum vanishes, then the momentum remains curl-free. Since viscosity spontaneously introduces a curl o the momentum, we cannot speak of irrotational flow of a viscous fluid.

For fluid flow there is a choice of coordinates to use. The customary coördinates — space and time — are known as Eulerian coordinates; they are not always the most suitable. Bobroff [3] has discussed three other sets of coordinates, some well suited for sma signal descriptions of electron beams. One that is widely used in fluid flow analysis is the Lagrangian set, which consists of time and the initial position of the particles, whereas the one that is bes suited for a description of fields is the Eulerian set. Various problems call for various sets of coordinates, and it is of interest to ex press the Manley-Rowe formulas in terms of these various sets.

In Sec. 6.2 (Eulerian Formulation), following the work of Wagne [110], we prove the formulas using Eulerian coordinates. The expressions for power flow are evaluated on a boundary that is fixed i space. Material may pass through this boundary, and in applying tl formulas this must be taken into account. The resulting formulas are suited for study of stability of hydrodynamic systems, but they are not as appropriate for problems in nonlinear acoustics, because in coupling to acoustic transducers the powers should be evaluated over a boundary that moves in space with the particles. The Lagrangian coordinates are thus the natural ones to use; in Sec. 6.2 (One-Dimensional Lagrangian Formulation) we prove the formulas for a one-dimensional acoustic system using Lagrangian coordinate

The expressions for power flow are different in the two cases.

Undoubtedly the Manley-Rowe formulas can be expressed in other sets of coordinates also.

Eulerian Formulation. The variables of the fluid-flow system are the particle density ρ, the particle velocity $\overline{v}$, the momentum per particle $\overline{p}$, and the pressure π. It is assumed that each of these can be defined as a continuous function of space and time, macroscopically.

These are related by the constitutive relations of the fluid. The velocity and momentum of each particle are related by an expression of the form

$$\overline{v} = \overline{v}(\overline{p}) \tag{6.33}$$

and therefore we can define a new variable, the kinetic energy per particle T, by the line integral

$$T = \int \overline{v}(\overline{p}) \cdot d\overline{p} \tag{6.34}$$

Thus, any variation of T is related to the variation of $\overline{p}$

$$\delta T = \overline{v} \cdot \delta\overline{p} \tag{6.35}$$

and in particular

$$\nabla T = (\overline{v} \cdot \nabla)\overline{p} + \overline{v} \times (\nabla \times \overline{p}) \tag{6.36}$$

Normally the relation of Eq. 6.33 is

$$\overline{p} = m\overline{v} \tag{6.37}$$

where m is the mass of the particle, but for relativistic flow this is modified to the form of Eq. 4.41. There is no need to specify what the relation of Eq. 6.33 is, so we do not.

Another constitutive relation is that between the pressure and the particle density

$$\pi = \pi(\rho) \tag{6.38}$$

from which we can calculate the intrinsic energy per particle by the integral

$$\Upsilon = \int \frac{\pi}{\rho^2}\, d\rho \tag{6.39}$$

so that

$$\nabla\left(\Upsilon + \frac{\pi}{\rho}\right) = \frac{\nabla\pi}{\rho} \tag{6.40}$$

where $(\Upsilon + \frac{\pi}{\rho})$ is the enthalpy per particle.

One of the equations of motion is the force equation,

$$\begin{aligned} \frac{D\overline{p}}{Dt} &= \dot{\overline{p}} + (v \cdot \nabla)\overline{p} \\ &= \text{Force per particle} \\ &= -\frac{\nabla\pi}{\rho} + \overline{F}_v \end{aligned} \tag{6.41}$$

where D/Dt is the substantive derivative (the time derivative taken in a reference frame moving along with the particles), and where $\overline{F}_v$ is the viscous force. From Eqs. 6.36, 6.40, and 6.41 one obtains

$$\nabla(T + \Upsilon + \frac{\pi}{\rho}) = -\dot{\overline{p}} + \overline{v} \times (\nabla \times \overline{p}) + \overline{F}_v \tag{6.42}$$

which is an alternate form of the force equation.

For systems with no viscosity and irrotational motion, we set $\overline{F}_v$ and $\nabla \times \overline{p}$ to zero, obtaining

$$\nabla(T + \Upsilon + \frac{\pi}{\rho}) = -\dot{\overline{p}} \tag{6.43}$$

The other equation of motion is the continuity equation, or law of conservation of particles,

$$\nabla \cdot (\rho\overline{v}) = -\dot{\rho} \tag{6.44}$$

From these equations of motion and the definitions of the variables T and Υ one can prove the power-flow theorem

$$\frac{\partial}{\partial t}(\rho T + \rho\Upsilon) = (T + \Upsilon + \frac{\pi}{\rho})\dot{\rho} + (\rho\overline{v}) \cdot \dot{\overline{p}}$$

$$= -\nabla \cdot (\rho\overline{v})(T + \Upsilon + \frac{\pi}{\rho}) \tag{6.45}$$

This is interpreted by identifying $\rho T + \rho\Upsilon$ with the energy density, and $(\rho\overline{v})(T + \Upsilon + \frac{\pi}{\rho})$ with the power-flow vector $\overline{F}$. The system is lossless; the energy density can increase only insofar as the power-flow vector has a nonzero divergence. The energy density is the sum of kinetic energy and intrinsic energy per unit volume, and the power flow vector has three terms: the term $\rho\overline{v}T$ is transport of kinetic energy by flow of particles; the term $\rho\overline{v}\Upsilon$ is transport of intrinsic energy by flow of particles; and the final term, $\pi\overline{v}$, represents a flow of power caused by movement of particles overcoming a pressure force.

Observe that the power-flow vector $\overline{F} = \rho\overline{v}T + \rho\overline{v}\Upsilon + \pi\overline{v}$ is not written, except for the last term, as a product of two variables. The electrical power-flow vector $\overline{E} \times \overline{H}$ is written as a product, and so the concept of "power at frequency ω_a" is very useful, since it is clear how to split apart the power-flow vector into two factors. The fluid-flow terms can be split apart more than one way; one of the results of our analysis is to give physical significance to one of these ways.

We may derive an expression similar to Eq. 6.45 involving the Fourier coefficients of the variables. We write the variables T, Υ, $\frac{\pi}{\rho}$, $\overline{p}$, ρ, and $(\rho\overline{v})$ in multiple Fourier series as in Eq. 3.15. Then Eq. 6.43 implies a similar relationship among the Fourier coefficients,

$$\nabla \left[T_a{}^* + \Upsilon_a{}^* + \left(\frac{\pi}{\rho} \right)_a{}^* \right] = j\omega_a \overline{p}_a{}^* \tag{6.46}$$

and Eq. 6.44 implies

$$\nabla \cdot (\rho \overline{v})_a = -j\omega_a \rho_a \tag{6.47}$$

If we multiply Eq. 6.47 by $\left[T_a{}^* + \Upsilon_a{}^* + \left(\frac{\pi}{\rho} \right)_a{}^* \right]$, and dot-multiply Eq. 6.46 by $(\rho\overline{v})_a$, and add, we find

$$j\omega_a \left\{ \left[T_a{}^* + \Upsilon_a{}^* + \left(\frac{\pi}{\rho} \right)_a{}^* \right] \rho_a - (\rho\overline{v})_a \cdot \overline{p}_a{}^* \right\}$$
$$= -\nabla \cdot \left\{ (\rho\overline{v})_a \left[T_a{}^* + \Upsilon_a{}^* + \left(\frac{\pi}{\rho} \right)_a{}^* \right] \right\} \tag{6.48}$$

The Manley-Rowe formulas now follow easily. If we consider $\rho T + \rho \Upsilon$ to be an energy state function depending on ρ and $\overline{p}$, the Manley-Rowe formulas are written in the form of Eq. 6.2 with the powers P_a given by

$$P_a = \tfrac{1}{2} \text{Re } j\omega_a \left\{ \left[T_a{}^* + \Upsilon_a{}^* + \left(\frac{\pi}{\rho} \right)_a{}^* \right] \rho_a + (\rho\overline{v})_a{}^* \cdot \overline{p}_a \right\}$$
$$= \tfrac{1}{2} \text{Re } j\omega_a \left\{ \left[T_a{}^* + \Upsilon_a{}^* + \left(\frac{\pi}{\rho} \right)_a{}^* \right] \rho_a - (\rho\overline{v})_a \cdot \overline{p}_a{}^* \right\} \tag{6.49}$$

or, by Eq. 6.48,

$$P_a = -\nabla \cdot \overline{F}_a \tag{6.50}$$

where the vector $\overline{F}_a$ is the power-flow vector at frequency ω_a,

$$\overline{F}_a = \tfrac{1}{2} \text{Re } (\rho\overline{v})_a \left[T_a{}^* + \Upsilon_a{}^* + \left(\frac{\pi}{\rho} \right)_a{}^* \right] \tag{6.51}$$

Thus the formulas can be written in the form of Eq. 6.4, and the powers can be evaluated on the boundary of the system.

One-Dimensional Lagrangian Formulation. In a one-dimensional system the variables depend on only one coordinate, x. At some initial time, t_0, we label each particle with its position x_0, and as the fluid moves, the particles retain a knowledge of their initial position, x_0, even though they move away from it. The variables of the system, ρ, v, ρv, T, Υ, etc., can be expressed either in terms of actual position x and time t, (the Eulerian set of coordinates) or in terms of the initial position x_0 of the particle and time t (the Lagrangian set). For notation, we denote partial derivatives with respect to the Eulerian coordinates by the ∂ symbol, and partial derivatives with respect to the Lagrangian coordinates by the D symbol. The time-differentiation D/Dt is the customary substantive derivative,

$$\frac{D}{Dt} = \frac{\partial}{\partial t} + v \cdot \nabla \tag{6.52}$$

We must express the equations of motion in terms of the Lagrangian coordinates. Eq. 6.41 becomes (neglecting viscosity)

$$\frac{Dp}{Dt} = -\frac{1}{\rho}\frac{\partial \pi}{\partial x} = -\frac{1}{\rho_0}\frac{D\pi}{Dx_0} \tag{6.53}$$

and Eq. 6.44 implies

$$\frac{D}{Dt}\left(\frac{1}{\rho}\right) = -\frac{1}{\rho^2}\frac{D\rho}{Dt}$$
$$= \frac{1}{\rho_0}\frac{Dv}{Dx_0} \tag{6.54}$$

where ρ_0, the initial particle density, is independent of time in the Lagrangian set of coordinates.

The kinetic energy per particle, T, is a function of p given by Eq. 6.34, and we consider the intrinsic energy per particle to be given by

$$\Upsilon\left(\frac{1}{\rho}\right) = -\int \pi \, d\left(\frac{1}{\rho}\right) \tag{6.55}$$

which is equivalent to Eq. 6.39, except that it is now a function of $(1/\rho)$, instead of ρ.

With the aid of Eqs. 6.53 and 6.54 we can compute the power-flow theorem

$$\frac{D}{Dt}\rho_0(T+\Upsilon) = \rho_0 v\frac{Dp}{Dt} - \rho_0\pi\frac{D}{Dt}\left(\frac{1}{\rho}\right)$$
$$= -v\frac{D\pi}{Dx_0} - \pi\frac{Dv}{Dx_0}$$
$$= -\frac{D}{Dx_0}(\pi v) \tag{6.56}$$

which is interpreted by identifying $\rho_0(T+\Upsilon)$ with the energy density, and πv with the power-flow, using Lagrangian coordinates.

We now express each of the variables in a multiple Fourier series where the Fourier coefficients depend on x_0, and rewrite Eqs. 6.53 and 6.54 as

$$j\omega_\alpha \rho_0 p_\alpha^* = \frac{D}{Dx_0}\pi_\alpha^* \tag{6.57}$$

and

$$j\omega_\alpha \rho_0\left(\frac{1}{\rho}\right)_\alpha = \frac{D}{Dx_0}v_\alpha \tag{6.58}$$

If the energy function $\rho_0(T+\Upsilon)$, as a function of p and $\frac{1}{\rho}$, is used to generate Manley-Rowe formulas, they are written in the form of Eq. 6.2 with the expression for P_α given by

$$P_\alpha = \tfrac{1}{2}\mathrm{Re}\; j\omega_\alpha \rho_0 \left[v_\alpha^* p_\alpha - \pi_\alpha^* \left(\frac{1}{\rho}\right)_\alpha \right]$$

$$= -\tfrac{1}{2}\mathrm{Re}\; j\omega_\alpha \rho_0 \left[v_\alpha p_\alpha^* + \pi_\alpha^* \left(\frac{1}{\rho}\right)_\alpha \right] \quad (6.59)$$

which is, in view of Eqs. 6.57 and 6.58,

$$P_\alpha = -\frac{D}{Dx_0} F_\alpha \quad (6.60)$$

where F_α is the component of power-flow πv caused by frequency ω_α,

$$F_\alpha = \tfrac{1}{2}\mathrm{Re}\; \pi_\alpha^* v_\alpha \quad (6.61)$$

The Manley-Rowe formulas of Eqs. 6.2 can now be integrated over the length of the system, and in view of Eq. 6.60 the power can be evaluated at the ends of the system, as the power input at frequency ω_α. Notice that F_α is evaluated by following along the particle at the end of the system, and noting its pressure and velocity as a function of time. Alternatively, if there is an acoustic transducer at the end of the system, the power may be evaluated as the force times the velocity of the diaphragm of the transducer. This Lagrangian formulation is therefore well suited for nonlinear one-dimensional acoustical systems.

We have attempted, without success, to find Manley-Rowe formulas for a three-dimensional acoustic system, using Lagrangian coordinates. It would be of interest to know under what conditions such formulas exist.

6.3 Irrotational Electron Beam

In this section, we show that an irrotational electron cloud in arbitrary electromagnetic fields obeys the Manley-Rowe formulas. One device this treatment applies to is the longitudinal electron beam with heavy pumping and small signal levels; this was discussed by Haus [36], who derived the Manley-Rowe formulas. His result is generalized here to irrotational beams and arbitrary signal levels.

The concept of "irrotational flow" is based on a well-known theorem proved in Appendix F, which states that if a beam starts out with zero curl of the momentum (with electromagnetic fields, this is the "canonical" momentum [30, Sec. 2.6]), it continues to move with zero curl [28]. Some practical conditions under which this holds [28] are discussed in Appendix F.

This system is, in a sense, a combination of the previous two systems, in which the particles have a charge q. For simplicity we neglect pressure effects, although they can be included very easily [110]. Although we speak as though the particles were electrons, the analysis clearly applies to any charged particles. In addition, it can be extended very easily to a system with two or more species of particles, each with its own mass and charge, provided each individually satisfies its own continuity equation, and provided they interact only through long range electromagnetic forces. Thus, in particular, macroscopic plasma models without viscosity and collisions might be included.

Our result is an extension of Chu's Kinetic Power Theorem [15], which is known to hold for longitudinal [41] and irrotational [48] electron beams. The gist of this theorem, in our present notation, is that for small signals each P_α individually vanishes. It is a small-signal power theorem, appropriate for small variations of a time-invariant device, with no frequency conversion taking place. It breaks down, however, when the disturbances are so large that non-linear effects occur. The Manley-Rowe formulas then express the constraints that remain in the nonlinear case.

We use Eulerian coordinates to describe the system, even though many electron beam analyses are done using other sets [3]. The fact that small-signal power theorems are available in other sets of coordinates [39, 100, 4] leads one to suspect that the Manley-Rowe formulas can also be expressed in these other coordinates, although we do not discuss them here.

The variables of this system are the electromagnetic field quantities $\overline{E}$, $\overline{H}$, $\overline{D}$, and $\overline{B}$, and the particle variables ρ, $\overline{v}$, and $\overline{p}$. The field variables $\overline{E}$ and $\overline{B}$ are related by two of Maxwell's equations, Eqs. 6.18 and 6.21; therefore, we may define, in the usual way [30, Sec. 1.5], the vector and scalar potentials $\overline{A}$ and Φ so that

$$\overline{B} = \nabla \times \overline{A} \tag{6.62}$$

and

$$\overline{E} = -\nabla\Phi - \dot{\overline{A}} \tag{6.63}$$

The momentum per particle, $\overline{p}$, is the canonical momentum [30, Sec. 2.6], which is, nonrelativistically,

$$\overline{p} = m\overline{v} + q\overline{A} \tag{6.64}$$

where m is the mass of each particle. The relativistic form for $\overline{p}$ is

$$\overline{p} = \frac{m\overline{v}}{\sqrt{1 - (v/c)^2}} + q\overline{A} \tag{6.65}$$

where m is the rest mass of the particles. In each case, the velocity is a function of $(\overline{p} - q\overline{A})$, and the kinetic energy per particle is

$$T = \int \overline{v}(\overline{p} - q\overline{A}) \cdot d(\overline{p} - q\overline{A}) \tag{6.66}$$

so that any variation δT of the kinetic energy is related to variations of $\overline{p}$ and $\overline{A}$

$$\delta T = \overline{v} \cdot \delta\overline{p} - q\overline{v} \cdot \delta\overline{A} \tag{6.67}$$

and in particular

$$\nabla T = (\overline{v} \cdot \nabla)\overline{p} - q(\overline{v} \cdot \nabla)\overline{A} + \overline{v} \times (\nabla \times \overline{p}) - q\overline{v} \times (\nabla \times \overline{A}) \tag{6.68}$$

The customary constitutive relations for the electromagnetic variables are those for free space, Eqs. 6.13 and 6.14. However, to allow for the possible presence of nonlinear, dispersion-free, anisotropic, inhomogeneous, lossless, reciprocal, stationary media in some parts of the system, we merely require relations of the form discussed in Sec. 6.1, so that the electric and magnetic energy densities W_e and W_m can be defined as in Eqs. 6.12 and 6.10.

The equations of motion include Maxwell's equations

$$\nabla \cdot \overline{B} = 0 \tag{6.69}$$

$$\nabla \cdot \overline{D} = q\rho \tag{6.70}$$

$$\nabla \times \overline{H} = q\rho\overline{v} + \dot{\overline{D}} \tag{6.71}$$

and

$$\nabla \times \overline{E} = -\dot{\overline{B}} \tag{6.72}$$

where we have identified the currents and charges as those of the electrons.

The mechanical equations of motion include the force equation,

$$\begin{aligned}\frac{D}{Dt}(\overline{p} - q\overline{A}) &= \dot{\overline{p}} - q\dot{\overline{A}} + (\overline{v} \cdot \nabla)\overline{p} - q(\overline{v} \cdot \nabla)\overline{A} \\ &= \text{Force per particle} \\ &= q\overline{E} + q\overline{v} \times \overline{B}\end{aligned} \tag{6.73}$$

which, with the aid of Eqs. 6.62 and 6. 68, becomes

$$\nabla T = -\dot{\overline{p}} + q\dot{\overline{A}} + q\overline{E} + \overline{v} \times (\nabla \times \overline{p}) \tag{6.74}$$

If the motion is irrotational, we set $\nabla \times \overline{p} = 0$, to obtain

$$\nabla T = -\dot{\overline{p}} + q\dot{\overline{A}} + q\overline{E} \tag{6.75}$$

The other mechanical equation of motion is the continuity equation,

$$\nabla \cdot (\rho\overline{v}) = -\dot{\rho} \tag{6.76}$$

From these equations of motion and the definitions of the energy state functions one can derive a power-flow theorem

$$\begin{aligned}\frac{\partial}{\partial t}(W_e + W_m + \rho T) &= T\dot{\rho} + (\rho\overline{v}) \cdot \dot{\overline{p}} - (q\rho\overline{v}) \cdot \dot{\overline{A}} + \overline{E} \cdot \dot{\overline{D}} + \overline{H} \cdot \dot{\overline{B}} \\ &= -\nabla \cdot (\overline{E} \times \overline{H} + \rho\overline{v}T)\end{aligned} \tag{6.77}$$

We interpret this by identifying $W_e + W_m + \rho T$ with the stored energy density, and $\overline{E} \times \overline{H} + \rho\overline{v}T$ with the power-flow vector $\overline{F}$.

In the same way that Eqs. 6.25 and 6.48 were derived, we can find a formula similar to Eq. 6.77, involving Fourier coefficients of the variables. It is

$$j\omega_a \left[(\rho\overline{v})_a^* \cdot \overline{p}_a - T_a \rho_a^* - (q\rho\overline{v})_a^* \cdot \overline{A}_a - \overline{E}_a \cdot \overline{D}_a^* + \overline{H}_a^* \cdot \overline{B}_a \right]$$

$$= - \nabla \cdot \left[\overline{E}_a \times \overline{H}_a^* + (\rho\overline{v})_a^* \, T_a \right] \qquad (6.78)$$

The Manley-Rowe formulas are now proved using $W_e + W_m + \rho T$ as the required energy state function, as a function of the variables $\overline{D}$, $\overline{B}$, ρ, and $(\overline{p} - q\overline{A})$. The expression for P_a is simply a combination of the expressions obtained in Secs. 6.1 (Manley-Rowe Formulas) and 6.2 (Eulerian Formulation):

$$P_a = - \nabla \cdot \overline{F}_a \qquad (6.79)$$

where

$$\overline{F}_a = \tfrac{1}{2}\text{Re}\left[\overline{E}_a \times \overline{H}_a^* + (\rho\overline{v})_a^* T_a \right] \qquad (6.80)$$

is the contribution to time-average power-flow vector $\overline{F} = \overline{E} \times \overline{H} + \rho$ caused by frequency ω_a. Because the Manley-Rowe formulas then are written in the form of Eq. 6.4, the powers can be evaluated at the boundary of the system.

6.4 Gyromagnetic Medium

We shall now show that a gyromagnetic medium obeys the Manley-Rowe formulas. The case of most practical interest is that of large pump and small signals, for which Haus [35] has proved the formulas. The more general case was first done by McWhorter [64], who used the Hamilton's principle approach to determine the correct variables to use. More precisely, McWhorter used the Hamiltonian state function, instead of the Lagrangian, with Hamilton's canonical equations as the equations of motion. Because the Hamiltonian density and the Lagrangian density are related by a Legendre transformation, the discussion of Sec. 4.1 shows that the same P_a is obtained with each. The important contribution was recognition of the appropriate variables. When these are known, the formulas can be proved directly from the equations of motion, as we show here.

One often uses the equations of motion of a gyromagnetic medium to describe ferrites [2, 102], if losses are neglected. Although the fir microwave parametric amplifiers were made from ferrites [103, 114] not until 1960 was it known that the gyromagnetic model obeys the ful Manley-Rowe formulas.

The variables of the gyromagnetic medium are the electromagnetic field quantities $\overline{E}$, $\overline{H}$, $\overline{D}$, and $\overline{B}$, and the magnetization vector $\overline{M}$ The constitutive relation between $\overline{D}$ and $\overline{E}$ is assumed to be

$$\overline{D} = \epsilon \overline{E} \tag{6.81}$$

where ϵ is the dielectric constant of the material. We can equally well assume that the relation between $\overline{D}$ and $\overline{E}$ is single-valued, nonlinear, dispersion-free, anisotropic, inhomogeneous, and reciprocal, as in Sec. 6.1. The important point is that the electric energy density

$$W_e(\overline{D}, \overline{r}) = \int \overline{E}(\overline{D}, \overline{r}) \cdot d\overline{D} \tag{6.82}$$

be defined.

The constitutive relation between the magnetic vectors, on the other hand, is (using, as usual, M.K.S. units)

$$\overline{B} = \mu_0(\overline{H} + \overline{M}) \tag{6.83}$$

where μ_0 is the permeability of free space.

The equations of motion include the four Maxwell's equations,

$$\nabla \cdot \overline{B} = 0 \tag{6.84}$$

$$\nabla \cdot \overline{D} = 0 \tag{6.85}$$

$$\nabla \times \overline{H} = \dot{\overline{D}} \tag{6.86}$$

and

$$\nabla \times \overline{E} = -\dot{\overline{B}} \tag{6.87}$$

and the magnetization equation,

$$\dot{\overline{M}} = -\gamma(\overline{M} \times \overline{B}) \tag{6.88}$$

where γ is a constant. Usually this relation is written with $\overline{B}$ replaced by $\overline{H}$, but by Eq. 6.83 we see that to write it using $\overline{B}$ we need merely redefine the constant γ.

It is clear from Eq. 6.88 that the magnitude of $\overline{M}$ is constant, for

$$\frac{\partial}{\partial t}(M^2) = 2\overline{M} \cdot \dot{\overline{M}} = 0 \tag{6.89}$$

Thus only two variables are necessary to determine $\overline{M}$, not three. The recognition of the proper variables to use [64] is the key to proving the Manley-Rowe formulas. We use the component of $\overline{M}$ in an arbitrary direction (call it the z-direction), M_z, and the angle ϕ made by the projection of $\overline{M}$ in the x-y plane. Thus

$$M_x = \sqrt{M^2 - M_z^2} \cos\phi \tag{6.90}$$

$$M_y = \sqrt{M^2 - M_z^2} \sin\phi \tag{6.91}$$

and

$$M_z = M_z \tag{6.92}$$

We now express Eq. 6.88 in terms of M_z and ϕ. First, we observe from Eqs. 6.90 and 6.91 that

$$\overline{B} \cdot \frac{\partial \overline{M}}{\partial \phi} = B_x \frac{\partial M_x}{\partial \phi} + B_y \frac{\partial M_y}{\partial \phi}$$
$$= M_x B_y - M_y B_x$$
$$= (\overline{M} \times \overline{B})_z \tag{6.93}$$

and therefore, from Eq. 6.88,

$$\dot{M}_z = -\gamma \overline{B} \cdot \frac{\partial \overline{M}}{\partial \phi} \tag{6.94}$$

Also, from Eq. 6.88,

$$\overline{B} \cdot \dot{\overline{M}} = 0 = \overline{B} \cdot \frac{\partial \overline{M}}{\partial \phi} \dot{\phi} + \overline{B} \cdot \frac{\partial \overline{M}}{\partial M_z} \dot{M}_z \tag{6.95}$$

and therefore, by Eq. 6.94,

$$\dot{\phi} = \gamma \overline{B} \cdot \frac{\partial \overline{M}}{\partial M_z} \tag{6.96}$$

Equations 6.94 and 6.96 are the equations of motion expressed in terms of the variables ϕ and M_z.

In proving the Manley-Rowe formulas, we use the state function

$$U = W_e(\overline{D}, \overline{r}) + \frac{B^2}{2\mu_0} - \overline{B} \cdot \overline{M} \tag{6.97}$$

which is a function of $\overline{D}$, $\overline{B}$, ϕ, and M_z. It is clear from Eqs. 6.94 and 6.96 that

$$\frac{\partial U}{\partial \phi} = \frac{\dot{M}_z}{\gamma} \tag{6.98}$$

and

$$\frac{\partial U}{\partial M_z} = -\frac{\dot{\phi}}{\gamma} \tag{6.99}$$

The Manley-Rowe formulas generated by U are written in the form of Eq. 6.2 with

$$P_a = \tfrac{1}{2}\mathrm{Re}\; j\omega_a \Big[\overline{E}_a^* \cdot \overline{D}_a + \frac{\overline{B}_a^*}{\mu_0} \cdot \overline{B}_a - \overline{M}_a^* \cdot \overline{B}_a$$
$$+ \left(\frac{\partial U}{\partial \phi}\right)_a^* \phi_a + \left(\frac{\partial U}{\partial M_z}\right)_a^* (M_z)_a \Big]$$
$$= \tfrac{1}{2}\mathrm{Re}\; j\omega_a \Big[\overline{E}_a^* \cdot \overline{D}_a + \overline{H}_a^* \cdot \overline{B}_a \Big]$$
$$= -\nabla \cdot \overline{F}_a \tag{6.100}$$

where

$$F_\alpha = \tfrac{1}{2}\mathrm{Re}\,(\overline{E}_\alpha \times \overline{H}_\alpha{}^*) \tag{6.101}$$

where we have used Eqs. 6.86, 6.87, 6.98, and 6.99, written in terms of the Fourier coefficients alone. The Manley-Rowe formulas can now be integrated over the system and put in the form of Eq. 6.4. The power quantities are the ordinary electromagnetic power flows at each frequency.

6.5 Hamilton's Principle

The distributed systems discussed in Secs. 6.1, 6.2, and 6.3 are all so simple that the Manley-Rowe formulas can be proved directly from the equations of motion. This cannot be done easily for all distributed systems, however, and a systematic procedure for obtaining the formulas is helpful. Such a method is available for systems that obey Hamilton's principle; in this section we demonstrate that Hamilton's principle leads to Manley-Rowe formulas with the powers P_α automatically in the form of divergences. This technique is applied to physical systems in Secs. 6.6 and 6.7.

Hamilton's principle for distributed systems [30, Ch. 11; 70, Part I, Ch. 3, Sec. 3. 4] is not as well known as the corresponding principle for lumped systems. It is a cumbersome way of obtaining the equations of motion, and invariably is stated only after the equations of motion are known. Its major uses in the past have been as a prelude to extending known equations of motion, either quantum-mechanically or relativistically. We use it here to generate Manley-Rowe formulas; a similar use in the past has been to generate, in roughly the same way, small-signal power theorems [100, 57].

For most systems the kinematic constraints are nonholonomic[30, Sec.1.3], and either potentials must be defined, or Lagrange multipliers used. There is no assurance that these physically unmeasurable quantities do not appear in the final expressions for the P_α.

The full Hamilton's principle method, with several variables in three dimensions, is complicated, so we give first, in Sec. 6.5 (A Simple Case), a simple derivation with only one variable and only one spacial dimension. The concepts are all illustrated by this simple case; the extension to any number of variables, and explicit space and time variation of the Lagrangian, in three dimensions, is straightforward, and is done in Sec. 6.5 (Extension).

Although we use Eulerian coordinates exclusively, there is no reason why Hamilton's principle cannot be formulated [100] in other coordinates [3].

A Simple Case. In distributed systems the energy state functions are densities; the Lagrangian density is used in Hamilton's principle. One of the variables of the system, η, together with its space and time derivatives, defines the Lagrangian

$$L = L(\eta, \dot{\eta}, \eta_x) \tag{6.102}$$

where we denote, temporarily, $\partial\eta/\partial x$ by η_x. Hamilton's principle states that the integral of L over both time and space is stationary with respect to a variation of η that vanishes on the boundary of the system and at the initial and final instants, or

$$\delta\int\int L \, dt \, dx = 0 \tag{6.103}$$

We introduce here some convenient notation. We need symbols for four types of differentiation:

(1) Regular partial differentiation of a variable or function with respect to time or space coordinates, as $\partial/\partial x$, $\partial/\partial y$, $\partial/\partial z$, and $\partial/\partial t$. The dot notation is reserved for this partial derivative with respect to time, and the vector operator ∇ is used for these space partial derivatives where convenient.

(2) In lumped systems, where t is the only independent coordinate, we have the total derivative d/dt.

(3) In fluid work, we use the substantive derivative $D/Dt = \partial/\partial t + (\overline{v} \cdot \nabla)$, where $\overline{v}$ is the particle velocity, to indicate time differentiation in a reference frame moving along with the particle.

(4) We need notation to indicate partial differentiation of an energy state function with respect to one of its variables, keeping the other variables fixed. Generally, such functions depend not only on the variables of the system, but on the independent coordinates (such as time) as well. Thus the usual ∂ notation is ambiguous. We denote such differentiation by Δ; for example, $\Delta/\Delta\eta$ means the partial derivative with respect to η, keeping the other variables on which the function depends constant. Thus the two quantities $\Delta L/\Delta t$ and $\dot{L} = \partial L/\partial t$ are quite different.

Using this notation, any variation in L is, to first order,

$$\delta L = \frac{\Delta L}{\Delta \eta} \, \delta\eta + \frac{\Delta L}{\Delta \dot{\eta}} \, \delta\dot{\eta} + \frac{\Delta L}{\Delta \eta_x} \, \delta\eta_x \tag{6.104}$$

One more convenient notation is the symbol $\doteq$ which will mean that the difference between the quantities it separates integrates out to zero. Thus, for example, for any scalar ξ,

$$\xi\delta\dot{\eta} \doteq -\dot{\xi}\delta\eta \tag{6.105}$$

since

$$\int_{t_1}^{t_2} (\xi\delta\dot{\eta} + \dot{\xi}\delta\eta) \, dt = \int_{t_1}^{t_2} \frac{d}{dt}(\xi\delta\eta) \, dt \tag{6.106}$$

vanishes, inasmuch as $\delta\eta$ vanishes at times t_1 and t_2. Similarly,

$$\nabla \cdot (\bar{\xi}\delta\eta) \doteq 0 \tag{6.107}$$

for any vector ξ.

Using this notation, Hamilton's principle states that

$$\delta L \doteq 0 \tag{6.108}$$

or,

$$0 \doteq \delta L = \frac{\Delta L}{\Delta \eta}\,\delta\eta + \frac{\Delta L}{\Delta \dot{\eta}}\,\delta\dot{\eta} + \frac{\Delta L}{\Delta \eta_x}\,\delta\eta_x$$

$$\doteq \frac{\Delta L}{\Delta \eta}\,\delta\eta - \frac{\partial}{\partial t}\frac{\Delta L}{\Delta \dot{\eta}}\,\delta\eta - \frac{\partial}{\partial x}\frac{\Delta L}{\Delta \eta_x}\,\delta\eta \tag{6.109}$$

The quantity on the last line of Eq. 6.109 can integrate to zero only if, for all time and for all space, the "Euler-Lagrange" equation

$$\frac{\Delta L}{\Delta \eta} = \frac{\partial}{\partial t}\frac{\Delta L}{\Delta \dot{\eta}} + \frac{\partial}{\partial x}\frac{\Delta L}{\Delta \eta_x} \tag{6.110}$$

remains valid.

The Hamiltonian state function H is defined by the Legendre transformation

$$H = \frac{\Delta L}{\Delta \dot{\eta}}\,\dot{\eta} - L \tag{6.111}$$

and so, by using Eq. 6.110, we obtain

$$\dot{H} = \frac{\partial H}{\partial t} = \frac{\partial}{\partial t}\left(\frac{\Delta L}{\Delta \dot{\eta}}\,\dot{\eta}\right) - \frac{\partial L}{\partial t}$$

$$= \dot{\eta}\frac{\partial}{\partial t}\frac{\Delta L}{\Delta \dot{\eta}} + \frac{\Delta L}{\Delta \dot{\eta}}\,\ddot{\eta} - \frac{\Delta L}{\Delta \eta}\,\dot{\eta} - \frac{\Delta L}{\Delta \dot{\eta}}\,\ddot{\eta} - \frac{\Delta L}{\Delta \eta_x}\,\dot{\eta}_x$$

$$= -\frac{\partial}{\partial x}\left(\frac{\Delta L}{\Delta \eta_x}\,\dot{\eta}\right) \tag{6.112}$$

If the Hamiltonian is to be identified with the energy density (as it often can be), Eq. 6.112 is an expression of the law of conservation of energy, with

$$\frac{\Delta L}{\Delta \eta_x}\,\dot{\eta} \tag{6.113}$$

playing the role of the power-flow vector (x-component).

To derive frequency-power formulas of Type I, we use either L or H as the required state function. The formulas are in the form of Eq. 6.2, with the P_a given by

$$P_a = \tfrac{1}{2}\mathrm{Re}\, j\omega_a \left[\left(\frac{\Delta L}{\Delta \eta}\right)_a^* \eta_a + \left(\frac{\Delta L}{\Delta \dot{\eta}}\right)_a^* \dot{\eta}_a + \left(\frac{\Delta L}{\Delta \eta_x}\right)_a^* \eta_{xa} \right] \quad (6.114)$$

But because Eq. 6.110 predicts

$$\left(\frac{\Delta L}{\Delta \eta}\right)_a^* = -\, j\omega_a \left(\frac{\Delta L}{\Delta \dot{\eta}}\right)_a^* + \frac{\partial}{\partial x}\left(\frac{\Delta L}{\Delta \eta_x}\right)_a^* \quad (6.115)$$

Eq. 6.114 becomes

$$P_a = \tfrac{1}{2}\mathrm{Re}\, \frac{\partial}{\partial x}\left[\left(\frac{\Delta L}{\Delta \eta_x}\right)_a^* \dot{\eta}_a \right] \quad (6.116)$$

Thus each P_a is put into the form of a divergence, and the form is expected from a knowledge of the actual power-flow vector of Eq. 6.113.

We have seen in this simple case that Hamilton's principle leads to P_a in the form of a divergence. This is extended in the next section.

Extension. Here we extend the treatment of the previous section to describe systems with several variables, and with explicit space and time variations, in three dimensions. It is assumed there are a number of variables defined as a function of space $\bar{r}$ and time t. In case some of the variables are vectors, we can treat each component separately, or we can use vector notation. The Lagrangian depends on variables η_i and on their derivatives $\partial \eta_i / \partial t$ and $\partial \eta_i / \partial x_j$ as well as on space and time explicitly.

$$L = L\left(\eta_i,\ \dot{\eta}_i,\ \frac{\partial \eta_i}{\partial x_j},\ \bar{r},\ t\right) \quad (6.117)$$

where (as in the future) i is an index running over the variables, and j is an index running over the three directions in space.

Hamilton's principle states that the integral of L is stationary with respect to variations of the η_i variables that are consistent with a set of "kinematic constraints." These constraints are often obvious from geometrical considerations for lumped systems, but in distributed systems they are far from obvious; it is necessary to introduce them purposefully. Virtually all kinematic constraints for distributed systems are nonholonomic [30, Sec. 1.3]; they are taken account of by two methods.

One is the definition of potential functions. For example, suppose one constraint is $\nabla \times \bar{A} = 0$ for some vector variable $\bar{A}$. We can account for this constraint by defining a scalar function ψ such that $\bar{A} = \nabla\psi$. We then express the Lagrangian in terms of ψ and its four derivatives, eliminating $\bar{A}$ as a variable.

The other method is that of Lagrange multipliers. If the nonholonomic constraint is

$$f\left(\eta_i, \dot{\eta}_i, \frac{\partial \eta_i}{\partial x_j}, x_j, t\right) = 0 \tag{6.118}$$

we define a Lagrange multiplier λ and a new Lagrangian

$$L' = L + \lambda f \tag{6.119}$$

The two Lagrangians are, of course, equal numerically, but are quite different functional forms; thus, in general,

$$\frac{\Delta L}{\Delta \eta_i} \neq \frac{\Delta L'}{\Delta \eta_i} \tag{6.120}$$

etc.

If there are several nonholonomic constraints, we may use the potential method on some and the multiplier method on the rest, if that is convenient. Usually neither the potentials nor the Lagrange multipliers are physically measurable quantities, although sometimes a physical interpretation can be given to them.

After the kinematic constraints (both holonomic and nonholonomic) have been put into the Lagrangian, either by direct elimination of variables, introduction of potentials, or addition of Lagrange multiplier terms, we remain with a Lagrangian that is a function of some of the original η_i variables and their four first-order derivatives, some potentials and their four first-order derivatives, some Lagrange multipliers, and space and time coordinates. We consider the potentials and Lagrange multipliers to be bona fide variables from now on, so that L is a function of some η_i, their first-order derivatives, and space and time coordinates,

$$L = L\left(\eta_i, \dot{\eta}_i, \frac{\partial \eta_i}{\partial x_j}, x_j, t\right) \tag{6.121}$$

Using this Lagrangian, Hamilton's principle states that

$$\delta L \doteq 0 \tag{6.122}$$

or,

$$0 \doteq \delta L = \Sigma_i \left[\frac{\Delta L}{\Delta \eta_i}\,\delta\eta_i + \frac{\Delta L}{\Delta \dot{\eta}_i}\,\delta\dot{\eta}_i + \Sigma_j \frac{\Delta L}{\Delta \dfrac{\partial \eta_i}{\partial x_j}}\,\delta\left(\frac{\partial \eta_i}{\partial x_j}\right)\right]$$

$$= \Sigma_i \left[\frac{\Delta L}{\Delta \eta_i}\,\delta\eta_i + \frac{\Delta L}{\Delta \dot{\eta}_i}\,\frac{\partial}{\partial t}\,\delta\eta_i + \Sigma_j \frac{\Delta L}{\Delta \dfrac{\partial \eta_i}{\partial x_j}}\,\frac{\partial}{\partial x_j}\,\delta\eta_i\right]$$

$$\doteq \Sigma_i \left[\frac{\Delta L}{\Delta \eta_i} - \frac{\partial}{\partial t}\,\frac{\Delta L}{\Delta \dot{\eta}_i} - \Sigma_j \frac{\partial}{\partial x_j}\,\frac{\Delta L}{\Delta \dfrac{\partial \eta_i}{\partial x_j}}\right]\delta\eta_i \tag{6.123}$$

The expression in Eq. 6.123 vanishes, when integrated over space and time, if and only if, for all i,

$$\frac{\Delta L}{\Delta \eta_i} = \frac{\partial}{\partial t} \frac{\Delta L}{\Delta \dot{\eta}_i} + \Sigma_j \frac{\partial}{\partial x_j} \frac{\Delta L}{\Delta \frac{\partial \eta_i}{\partial x_j}} \tag{6.124}$$

for all time and at each point in space. Equations 6.124 are the Euler-Lagrange equations. The equations corresponding to the Lagrange multipliers are merely the kinematic constraints that the Lagrange multipliers were defined for.

If Hamilton's principle is correct, the Euler-Lagrange equations should predict the ordinary equations of motion, and vice versa. It is not a trivial task to show that this is true. Equations 6.124 contain potentials and Lagrange multipliers; these must be eliminated to derive the ordinary equations of motion, and must be defined in any attempt to show the generality of the Euler-Lagrange equations.

The Hamiltonian H is defined by the Legendre transformation

$$\mathrm{H} = \Sigma_i \frac{\Delta L}{\Delta \dot{\eta}_i} \dot{\eta}_i - L \tag{6.125}$$

and so by use of Eq. 6.124, the time-derivative of H is

$$\dot{\mathrm{H}} = \frac{\partial}{\partial t}\left[\Sigma_i \frac{\Delta L}{\Delta \dot{\eta}_i} \dot{\eta}_i\right] - \dot{L}$$

$$= \Sigma_i \left[\dot{\eta}_i \frac{\partial}{\partial t} \frac{\Delta L}{\Delta \dot{\eta}_i} + \frac{\Delta L}{\Delta \dot{\eta}_i} \ddot{\eta}_i - \frac{\Delta L}{\Delta \eta_i} \dot{\eta}_i - \frac{\Delta L}{\Delta \dot{\eta}_i} \ddot{\eta}_i - \Sigma_j \frac{\Delta L}{\Delta \frac{\partial \eta_i}{\partial x_j}} \frac{\partial \dot{\eta}_i}{\partial x_j}\right] - \frac{\Delta L}{\Delta t}$$

$$= \Sigma_i \left[- \dot{\eta}_i \Sigma_j \frac{\partial}{\partial x_j} \frac{\Delta L}{\Delta \frac{\partial \eta_i}{\partial x_j}} - \Sigma_j \frac{\Delta L}{\Delta \frac{\partial \eta_i}{\partial x_j}} \frac{\partial \dot{\eta}_i}{\partial x_j}\right] - \frac{\Delta L}{\Delta t}$$

$$= - \Sigma_i \Sigma_j \frac{\partial}{\partial x_j} \left(\frac{\Delta L}{\Delta \frac{\partial \eta_i}{\partial x_j}} \dot{\eta}_i\right) - \frac{\Delta L}{\Delta t} \tag{6.126}$$

so we can write

$$\dot{\mathrm{H}} + \nabla \cdot \overline{F} = - \frac{\Delta L}{\Delta t} \tag{6.127}$$

where the vector $\overline{F}$ has components

$$F_j = \Sigma_i \frac{\Delta L}{\Delta \frac{\partial \eta_i}{\partial x_j}} \dot{\eta}_i \tag{6.128}$$

If $\mathcal{H}$ is the internal energy density, then we interpret Eq. 6.127 as the law of conservation of energy, where $\overline{F}$ is the power-flow vector. More often, however, $\mathcal{H}$ is not the energy density, and one must use care in interpreting $\overline{F}$.

We can obtain formulas similar to Eq. 6.124 but involving the Fourier coefficients of the variables,

$$\left(\frac{\Delta L}{\Delta \eta_i}\right)^*_a = -j\omega_a \left(\frac{\Delta L}{\Delta \dot{\eta}_i}\right)^*_a + \Sigma_j \frac{\partial}{\partial x_j}\left(\frac{\Delta L}{\Delta \frac{\partial \eta_i}{\partial x_j}}\right)^*_a \tag{6.129}$$

The Lagrangian state function generates Manley-Rowe formulas of the form of Eq. 6.2, with the powers P_a given by

$$P_a = \tfrac{1}{2}\mathrm{Re}\,\Sigma_i j\omega_a \left[\left(\frac{\Delta L}{\Delta \eta_i}\right)^*_a (\eta_i)_a + \left(\frac{\Delta L}{\Delta \dot{\eta}_i}\right)^*_a (\dot{\eta}_i)_a + \Sigma_j \left(\frac{\Delta L}{\Delta \frac{\partial \eta_i}{\partial x_j}}\right)^*_a \frac{\partial(\eta_i)_a}{\partial x_j}\right] \tag{6.130}$$

which, with the aid of Eq. 6.129, can be easily put into the form

$$P_a = -\nabla\cdot\overline{F}_a \tag{6.131}$$

where the vector $\overline{F}_a$ has components

$$F_{aj} = \tfrac{1}{2}\mathrm{Re}\,\Sigma_i \left(\frac{\Delta L}{\Delta \frac{\partial \eta_i}{\partial x_j}}\right)^*_a (\dot{\eta}_i)_a \tag{6.132}$$

and is therefore the contribution to time-average value of $\overline{F}$ caused by frequency ω_a. The Manley-Rowe formulas can now be integrated to the form of Eq. 6.4, and the powers can be evaluated on the boundary of the system.

In general, the expressions for the $\overline{F}_a$ contain the Fourier coefficients of the physically unmeasurable potentials and Lagrange multipliers. Care must then be used in interpreting the Manley-Rowe formulas. The physical systems discussed in Secs. 6.1, 6.2, 6.3, and 6.4 obey Hamilton's principle, and it happens that the expressions for the $\overline{F}_a$ do not contain the potentials that must be introduced [78]. One should not conclude, however, that this happens in all systems; the physical systems discussed in Secs. 6.6 and 6.7 have Manley-Rowe formulas including physically unmeasurable terms.

6.6 Rotational Fluid Flow

Here we discuss, in detail, a physical system with rotational flow of material, with or without electric charges. The analysis, which can be specialized to simpler systems merely by eliminating the irrelevant terms, is for isentropic flow of a charged fluid, which is compressible with a scalar pressure, under the action of arbitrary electric and magnetic fields and a gravity potential.

This system is so complicated that the systematic procedure of Hamilton's principle actually simplifies the problem. In Sec. 6.6 (Constitutive Relations) we discuss the properties of the system. The equations of motion are given in the second part of Sec. 6.6, and the power-flow theorem is derived. The Lagrangian is given in the third part of this section, and in Sec. 6.6 (Hamilton's Principle) it is shown that the Euler-Lagrange equations coming from this Lagrangian are equivalent to the ordinary equations of motion. The Manley-Rowe formulas are given in the last portion of this section.

Free charges and currents are not included in this system, but in Sec. 6.7 the modifications that are necessary are discussed.

The Constitutive Relations. The electromagnetic variables are the electric field $\overline{E}$, the magnetic field $\overline{H}$, the electric displacement vector $\overline{D}$, and the magnetic induction vector $\overline{B}$. At the outset we recognize that $\overline{E}$ and $\overline{B}$ are related by two of Maxwell's equations

$$\nabla \cdot \overline{B} = 0 \tag{6.133}$$

and

$$\nabla \times \overline{E} = -\dot{\overline{B}} \tag{6.134}$$

so we can define [30, Sec. 1.5] the usual vector and scalar potentials $\overline{A}$ and Φ, so that

$$\overline{B} = \nabla \times \overline{A} \tag{6.135}$$

and

$$\overline{E} = -\nabla\Phi - \dot{\overline{A}} \tag{6.136}$$

The fluid variables are chosen on a "per-particle" basis. It is assumed that each can be defined macroscopically as a function of space $\overline{r}$ and time t. We let ρ be the number density of particles, π the pressure (a scalar), θ the temperature, s the entropy, $\overline{\iota} = \rho\overline{v}$ the particle current density, $\overline{p}$ the momentum per particle, and the gravitational force per particle is $-\nabla\Omega$, where Ω is the gravitational potential, assumed to be time-invariant.

The momentum $\overline{p}$ is, as in Sec. 6.3, the "canonical momentum," which is not physically measurable; however, $\overline{p} - q\overline{A}$ is physically measurable, where q is the charge per particle. We assume, as in Sec. 6.3, that $\overline{p} - q\overline{A}$ is a function of the velocity, so we can define a new variable, the kinetic coenergy per particle,

$$T' = \int (\overline{p} - q\overline{A}) \cdot d\overline{v} \tag{6.137}$$

We also define the "kinetic energy per particle" T by the Legendre transformation

$$T = (\overline{p} - q\overline{A}) \cdot \overline{v} - T' \tag{6.138}$$

In the theory of the motion of a single particle, T and T' play the role of energy state functions, but in our analysis they are considered as variables. We think of T' as a function jointly of the particle density ρ and the particle current density $\overline{\iota}$, rather than of the velocity $\overline{v}$.

The functional dependence of $\overline{p}$ is, nonrelativistically,

$$\overline{p} = q\overline{A} + m\overline{v} \tag{6.139}$$

so

$$T = \frac{1}{2m}(p - qA)^2 \tag{6.140}$$

where m is the mass per particle. The relativistic dependence is more complicated:

$$\overline{p} = q\overline{A} + \frac{mc\overline{v}}{\sqrt{c^2 - v^2}} \tag{6.141}$$

thus

$$T' = -mc(c^2 - v^2)^{\frac{1}{2}} \tag{6.142}$$

and

$$T = (m^2c^4 + (p - qA)^2c^2)^{\frac{1}{2}} \tag{6.143}$$

where m is the rest mass of the particle, and c is the velocity of light.

The electrical constitutive relations most often used are those for free space,

$$\overline{D} = \epsilon_0\overline{E} \tag{6.144}$$

and

$$\overline{B} = \mu_0\overline{H} \tag{6.145}$$

where ϵ_0 is the permittivity of free space, and μ_0 is the permeability of free space. However, to allow nonlinear, dispersion-free, anisotropic, inhomogeneous, reciprocal, stationary electromagnetic media to occupy some parts of our system, we merely require (as in Sec. 6.1) that $\overline{H}$ be a reciprocal function of $\overline{B}$ and spacial coordinates $\overline{r}$, and that $\overline{D}$ be a reciprocal function of $\overline{E}$ and $\overline{r}$, so that we can define the energy state functions

$$W_m(\overline{B}, \overline{r}) = \text{Magnetic energy density}$$

$$= \int \overline{H}(\overline{B}, \overline{r}) \cdot d\overline{B} \tag{6.146}$$

$$W_e'(\overline{E}, \overline{r}) = \text{Electric coenergy density}$$
$$= \int \overline{D}(\overline{E}, \overline{r}) \cdot d\overline{E} \tag{6.147}$$

and

$$W_e(\overline{D}, \overline{r}) = \text{Electric energy density}$$
$$= \overline{E} \cdot \overline{D} - W_e'(\overline{E}, \overline{r}) \tag{6.148}$$

The remaining constitutive relations are among the thermodynamic variables ρ, s, θ, and π. We consider pressure and temperature to be functions of ρ and s. On thermodynamic grounds we define the "internal energy per particle" or "intrinsic energy per particle" Υ by

$$\Upsilon(\rho, s) = \int \frac{\pi(\rho, s)}{\rho^2} d\rho + \int \theta(\rho, s)\, ds \tag{6.149}$$

Equations of Motion. We choose to treat only isentropic flow, by which we mean that the entropy of a given particle remains constant. This restriction is expressed as the equation of isentropy,

$$\frac{Ds}{Dt} = \dot{s} + \overline{v} \cdot \nabla s = 0 \tag{6.150}$$

where D/Dt indicates the substantive derivative, taken in a reference frame that at each instant moves with the particle.

The equations of motion include the force equation,

$$\frac{D(\overline{p} - q\overline{A})}{Dt} = \dot{\overline{p}} - q\dot{\overline{A}} + (\overline{v} \cdot \nabla)\overline{p} - q(\overline{v} \cdot \nabla)\overline{A}$$
$$= \text{Force per particle}$$
$$= -\frac{\nabla\pi}{\rho} + q\overline{E} + q\overline{v} \times \overline{B} - \nabla\Omega \tag{6.151}$$

We note from Eq. 6.138 and the preceding discussion that (just as in Sec. 6.3)

$$\nabla T = (\overline{v} \cdot \nabla)\overline{p} - q(\overline{v} \cdot \nabla)\overline{A} + \overline{v} \times (\nabla \times \overline{p}) - q\overline{v} \times (\nabla \times \overline{A}) \tag{6.152}$$

From Eq. 6.149 it follows that

$$\nabla(\Upsilon + \frac{\pi}{\rho}) = \frac{\nabla\pi}{\rho} + \theta\nabla s \tag{6.153}$$

so that with the aid of Eq. 6.135

$$\nabla(T + \Upsilon + \frac{\pi}{\rho} + \Omega) = -\dot{\overline{p}} + q\dot{\overline{A}} + q\overline{E} + \overline{v} \times (\nabla \times \overline{p}) + \theta\nabla s \tag{6.154}$$

which is an alternate form of the force equation. This is one of the equations of motion. In addition, there is the equation of isentropy, here multiplied by ρ,

$$\rho \dot{s} + \overline{\iota} \cdot \nabla s = 0 \tag{6.155}$$

We also have the equation of continuity, or particle conservation,

$$\nabla \cdot \overline{\iota} = -\dot{\rho} \tag{6.156}$$

and Maxwell's equations, Eq. 6.133, 6.134, and

$$\nabla \cdot \overline{D} = q\rho \tag{6.157}$$

$$\nabla \times \overline{H} = q\overline{\iota} + \dot{\overline{D}} \tag{6.158}$$

We have assumed that the only charges and currents are those caused by the charged particles and their motion; that is, we allow no "free" charges or currents, such as appear in magnetohydrodynamic systems. These are discussed in Sec. 6.7.

The power-flow theorem follows from the equations of motion just given. Dot-multiply Eq. 6.154 by $\overline{\iota}$, multiply Eq. 6.156 by $(T + \Upsilon + \frac{\pi}{\rho} + \Omega)$, dot-multiply Eq. 6.134 by $\overline{H}$, and dot-multiply Eq. 6.158 by $-\overline{E}$. Then add the four resulting equations, combining the left-hand sides into the divergence of the power-flow vector $\overline{F}$

$$\overline{F} = \overline{\iota}\,(T + \Upsilon + \frac{\pi}{\rho} + \Omega) + \overline{E} \times \overline{H} \tag{6.159}$$

The right-hand side is

$$-\frac{\partial}{\partial t}(W_e + W_m + \rho T + \rho \Upsilon + \rho \Omega) \tag{6.160}$$

The power-flow vector $\overline{F}$ is a sum of terms representing flow of kinetic, intrinsic, and gravitational energy of the particles, and the term $\pi\overline{v}$, which represents a flow of power caused by motion of particles in overcoming a pressure force, and finally the electromagnetic power-flow term. The expression for energy density, Eq. 6.160, includes electric, magnetic, kinetic, intrinsic, and gravitational energy densities.

We have now recognized the power-flow vector $\overline{F}$ of Eq. 6.159; when the Manley-Rowe formulas are obtained, we will try to interpret the formulas for $\overline{F}_\alpha$ in terms of Eq. 6.159.

The Lagrangian. The Lagrangian density L is formed by taking the kinetic coenergy density, and subtracting from it the gravitational potential energy density and the intrinsic energy density. Added to this is the Lagrangian for the electromagnetic field, with coupling terms. The result is

$$L = \rho T' - \rho \Upsilon - \rho \Omega + W_e' - W_m - q\rho\Phi + q\overline{\iota} \cdot \overline{A} \tag{6.161}$$

As it stands, the Lagrangian does not predict the correct equations of motion, because there are kinematic constraints. Two of the kinematic constraints, Eqs. 6.133 and 6.134, have already been used to define the potentials $\overline{A}$ and Φ. Another kinematic constraint is the continuity equation, Eq. 6.156, which we account for with a Lagrang multiplier, ζ. Another kinematic constraint is the equation of isen tropy, Eq. 6.155, for which we define the Lagrange multiplier β.

If these were the only kinematic constraints used, the resulting equations of motion would not be general, but instead would predict only a limited class of flows. One of the resulting Euler-Lagrange equations would be

$$\nabla\zeta = -\overline{p} + \beta\nabla s \tag{6.162}$$

and so, for uniform entropy, only irrotational flow would be predicted.

Early attempts to predict more general motion are those of Itô [45], Ziman [121], and Thellung [105]. These authors employed a Clebsch transformation [51, pp. 248-249] expressing the momentum $\overline{p}$ in the form

$$\overline{p} = -\nabla\phi - \chi\nabla\psi \tag{6.163}$$

Neglecting for the moment the electrical and thermodynamic variables, the Lagrangian is taken to be

$$\rho(\dot{\phi} + \chi\dot{\psi} - T - \Upsilon - \Omega) \tag{6.164}$$

and is therefore a function of ρ, ϕ, χ, and ψ, and their derivatives. The Euler-Lagrange equations are the force equation, the equation of continuity, and the two equations

$$\frac{D\chi}{Dt} = 0 \tag{6.165}$$

and

$$\frac{D\psi}{Dt} = 0 \tag{6.166}$$

This derivation, using the Clebsch transformation, has been extend to include electromagnetic forces [67] and relativistic motion [112]. I does not require the introduction of the continuity equation as a kinematic constraint, but the form of the Lagrangian and the potentials are not easily given a physical interpretation. We prefer to use an alternate method, which is similar analytically, but different in phil osophy.

This method, due to Lin [54, 92], shows clearly the particle-like basis of fluid mechanics. At some time t_0 we give to each particle of our system a label which is its position at that time, $\overline{a}$. The par ticle keeps this label throughout its travel, and consequently we can define a vector field $\overline{a}$ as a function of space $\overline{r}$ and time t. Opera tionally, we determine $\overline{a}$ corresponding to a given point $\overline{r}$ at a giv time t by looking at that point, at that time, and noting what partic is there. The value of $\overline{a}$ associated with that particle then is defin as $\overline{a}(\overline{r}, t)$.

Stated another way, we go to a point $\bar{r}$ and at time t we ask the particle passing by, "Where were you at time t_0?" The particle's answer defines the value of the vector field $\bar{a}$ at $\bar{r}$ and t.

The equation expressing the fact that each particle keeps the same value of $\bar{a}$ is

$$\frac{D\bar{a}}{Dt} = \dot{\bar{a}} + (\bar{v} \cdot \nabla)\bar{a} = 0 \tag{6.167}$$

or, when multiplied by ρ,

$$\rho\dot{\bar{a}} + (\bar{\iota} \cdot \nabla)\bar{a} = 0 \tag{6.168}$$

We use this as a kinematic constraint, to be handled with a Lagrange multiplier $\bar{\lambda}$.

The Lagrangian of Eq. 6.161 is then replaced by

$$\begin{aligned} L = {} & \rho T' - \rho\Upsilon - \rho\Omega + W_e' - W_m - q\rho\Phi + q\bar{\iota} \cdot \bar{A} \\ & - \zeta(\dot{\rho} + \nabla \cdot \bar{\iota}) - \beta(\rho\dot{s} + \bar{\iota} \cdot \nabla s) \\ & - \bar{\lambda} \cdot [\rho\dot{\bar{a}} + (\bar{\iota} \cdot \nabla)\bar{a}] \end{aligned} \tag{6.169}$$

which is considered a function of the variables ρ, $\bar{\iota}$, s, Φ, $\bar{a}$, ζ, $\bar{A}$, β, $\bar{\lambda}$, and their space and time derivatives, as well as space explicitly.

Hamilton's Principle. Hamilton's principle, Eq. 6.122, predicts Euler-Lagrange equations in the form of Eq. 6.124. We omit the details and write down the Euler-Lagrange equations. There is one such equation associated with each variable.

Corresponding to the variation in ρ we find

$$\dot{\zeta} = T + \Upsilon + \frac{\pi}{\rho} + \Omega + q\Phi + \bar{\lambda} \cdot \dot{\bar{a}} + \beta\dot{s} \tag{6.170}$$

the equation obtained by varying $\bar{\iota}$ is

$$\nabla\zeta = -\bar{p} + \Sigma_j \lambda_j \nabla a_j + \beta\nabla s \tag{6.171}$$

where, as in the following, Σ_j indicates a sum over the three directions in space; the variation in s yields

$$\rho\theta - \frac{\partial}{\partial t}(\rho\beta) - \nabla \cdot (\bar{\iota}\beta) = 0 \tag{6.172}$$

which, with the aid of Eq. 6.178, can be put into the form

$$\frac{D\beta}{Dt} = \dot{\beta} + \bar{v} \cdot \nabla\beta = \theta \tag{6.173}$$

the formula resulting from the variable $\bar{A}$ is

$$\nabla \times \bar{H} = q\bar{\iota} + \dot{\bar{D}} \tag{6.174}$$

associated with the potential Φ is the equation

$$\nabla \cdot \overline{D} = q\rho \tag{6.175}$$

the formula from the variation in $\overline{a}$ is

$$\frac{\partial}{\partial t}(\rho\lambda_j) + \nabla \cdot (\overline{\iota}\lambda_j) = 0 \tag{6.176}$$

or, in view of Eq. 6.178,

$$\frac{D\overline{\lambda}}{Dt} = \dot{\overline{\lambda}} + (\overline{v} \cdot \nabla)\overline{\lambda} = 0 \tag{6.177}$$

corresponding to the variations in the Lagrange multipliers, ζ, β, and $\overline{\lambda}$ we obtain merely the three kinematic constraints,

$$\dot{\rho} + \nabla \cdot \overline{\iota} = 0 \tag{6.178}$$

$$\rho\dot{s} + \overline{\iota} \cdot \nabla s = 0 \tag{6.179}$$

and

$$\rho\dot{\overline{a}} + (\overline{\iota} \cdot \nabla)\overline{a} = 0 \tag{6.180}$$

It is not obvious that these Euler-Lagrange equations are identical to the ordinary equations of motion. To show that they are, we must derive the ordinary equations of motion from the Euler-Lagrange equations, and vice versa.

It is convenient to introduce a lemma to help do this. Let us suppose that $\overline{a}$ and $\overline{\lambda}$ are any two vectors whose substantive derivatives vanish:

$$\dot{\overline{a}} + (\overline{v} \cdot \nabla)\overline{a} = 0 \tag{6.181}$$

and

$$\dot{\overline{\lambda}} + (\overline{v} \cdot \nabla)\overline{\lambda} = 0 \tag{6.182}$$

also let us suppose that s, β, and θ are scalar functions of space and time such that

$$\dot{s} + \overline{v} \cdot \nabla s = 0 \tag{6.183}$$

and

$$\dot{\beta} + \overline{v} \cdot \nabla\beta = \theta \tag{6.184}$$

Then it is identically true that

$$v \times \left[\nabla \times (\Sigma_j \lambda_j \nabla a_j) + \beta\nabla s)\right] + \theta\nabla s$$

$$= \frac{\partial}{\partial t}(\Sigma_j \lambda_j \nabla a_j + \beta\nabla s) - \nabla(\overline{\lambda} \cdot \dot{\overline{a}} + \beta\dot{s}) \tag{6.185}$$

We now show that the ordinary equations of motion can be derived from the Euler-Lagrange equations. The only problem is in deriving the force equation, Eq. 6.154. We recognize the left-hand side of Eq. 6.185 as

$$\bar{v} \times (\nabla \times \bar{p}) + \theta \nabla s \tag{6.186}$$

by use of the Euler-Lagrange Eq. 6.171. We now take the gradient of Eq. 6.170, and subtract from it the time-derivative of Eq. 6.171, to eliminate the Lagrange multiplier ζ. The result is

$$\nabla (T + \Upsilon + \frac{\pi}{\rho} + \Omega) = - q\nabla\Phi - \nabla(\bar{\lambda} \cdot \dot{\bar{a}} + \beta\dot{s}) - \dot{\bar{p}} + \frac{\partial}{\partial t} (\Sigma_j \lambda_j \nabla a_j + \beta \nabla s) \tag{6.187}$$

which, using Eq. 6.136 and the lemma just proved, reduces to

$$\nabla(T + \Upsilon + \frac{\pi}{\rho} + \Omega) = -\dot{\bar{p}} + q\dot{\bar{A}} + q\bar{E} + \bar{v} \times (\nabla \times \bar{p}) + \theta\nabla s \tag{6.188}$$

which is identical to the force equation, Eq. 6.154. We have thus shown that Hamilton's principle predicts the correct equations of motion.

The fact that the Euler-Lagrange equations predict the correct equations of motion does not imply that the converse is true — that is, that the Euler-Lagrange equations are general enough to predict all possible types of motion. This is shown by deriving the form of the Euler-Lagrange equations from the known equations of motion. The earliest formulations of the fluid-flow problem in Eulerian coordinates failed to predict the most general type of motion. Lin's method, which we have used here, does predict the most general motion; this was first shown by Serrin, in a letter [91] to Lin. We use here a method somewhat different from Serrin's.

The only Euler-Lagrange equations that we need verify are Eqs. 6.170, 6.171, 6.173, 6.177, and 6.180, since the others are identical to the equations of motion. Further, Eq. 6.180 is automatically satisfied if we define the vector field $\bar{a}(\bar{r}, t)$, as before, as the initial position vector of the particle at $\bar{r}$ at time t.

Equation 6.173 can be satisfied by defining β properly. Let β be the time-integral of the temperature of a given particle, starting at time t_0. Thus Eq. 6.173 holds, and $\beta(\bar{r}, t_0) = 0$.

Similarly, we can define a vector field $\bar{\lambda}$ so that Eq. 6.177 is satisfied. At time t_0, let $\bar{\lambda}$ be any vector with a curl identical to the curl of the momentum $\bar{p}$. For example, $\bar{\lambda}$ could be chosen equal to the momentum at time t_0. Now, at each point we assign the value of $\bar{\lambda}$ to the particle there, and let the particle carry this value as a label, just as it does the value of $\bar{a}$. From this definition it is clear that the substantive derivative of $\bar{\lambda}$ vanishes, and Eq. 6.177 holds.

Now we need only prove Eqs. 6.170 and 6.171. We use the lemma just proved above. We have chosen $\bar{a}$, $\bar{\lambda}$, β, and θ so that Eqs. 6.181, 6.182, 6.183, and 6.184 hold, so the result of Eq. 6.185 is true. We define a vector field $\bar{u}(\bar{r}, t)$ by

$$\bar{u} = \nabla \times (\bar{p} - \Sigma_j \lambda_j \nabla a_j - \beta \nabla s) \tag{6.189}$$

noting that at time t_0, $\overline{u}$ vanishes everywhere. From the force equation, Eq. 6.154,

$$\nabla \times \dot{\overline{p}} = \nabla \times [\overline{v} \times (\nabla \times \overline{p})] + \nabla \times (\theta \nabla s) \tag{6.190}$$

from Eq. 6.185 we can evaluate $\nabla \times (\overline{v} \times \overline{u})$ as

$$\nabla \times (\overline{v} \times \overline{u}) = \nabla \times [\overline{v} \times (\nabla \times \overline{p})] + \nabla \times (\theta \nabla s) - \frac{\partial}{\partial \iota}\left[\nabla \times (\Sigma_j \lambda_j \nabla a_j + \beta \nabla s)\right] \tag{6.191}$$

thus

$$\begin{aligned}\dot{\overline{u}} &= \nabla \times (\overline{v} \times \overline{u}) \\ &= (\overline{u} \cdot \nabla)\overline{v} - (\overline{v} \cdot \nabla)\overline{u} - \overline{u}(\nabla \cdot \overline{v})\end{aligned} \tag{6.192}$$

Equation 6.192 then shows that

$$\frac{D\overline{u}}{Dt} = (\overline{u} \cdot \nabla)\overline{v} - \overline{u}(\nabla \cdot \overline{v}) \tag{6.193}$$

With specified flow, this is a set of three coupled linear time-varyin total differential equations for the three components of $\overline{u}$; the solution is thus unique. A possible solution is the trivial one $\overline{u} = 0$, and because $\overline{u} = 0$ everywhere at time t_0, we see that this is therefore the solution. Thus

$$\overline{u} = \nabla \times (\overline{p} - \Sigma_j \lambda_j \nabla a_j - \beta \nabla s) = 0 \tag{6.194}$$

so that for some scalar $\upsilon(\overline{r}, t)$,

$$\Sigma_j \lambda_j \nabla a_j + \beta \nabla s - \overline{p} = \nabla \upsilon \tag{6.195}$$

In view of Eqs. 6.194 and 6.195 the lemma of Eq. 6.185 reads

$$\overline{v} \times (\nabla \times \overline{p}) + \theta \nabla s = \dot{\overline{p}} + \nabla \dot{\upsilon} - \nabla(\overline{\lambda} \cdot \dot{\overline{a}} + \beta \dot{s}) \tag{6.196}$$

so that the force equation, Eq. 6.154, can be altered to the form

$$\nabla(T + \Upsilon + \frac{\pi}{\rho} + \Omega + q\Phi) = \nabla \dot{\upsilon} - \nabla(\overline{\lambda} \cdot \dot{\overline{a}} + \beta \dot{s}) \tag{6.197}$$

But two scalars with equal gradients can differ only by a function of time, so for some $\nu(t)$,

$$T + \Upsilon + \frac{\pi}{\rho} + \Omega + q\Phi + \overline{\lambda} \cdot \dot{\overline{a}} + \beta \dot{s} = \dot{\upsilon} + \nu \tag{6.198}$$

We now define the scalar function of space and time $\zeta(\overline{r}, t)$ as

$$\zeta(\overline{r}, t) = \int_{t_0}^{t} \nu(t)dt + \upsilon(\overline{r}, t) \tag{6.199}$$

so that

$$\mathring{\zeta} = \nu + \mathring{\upsilon}$$

$$= T + \Upsilon + \frac{\pi}{\rho} + \Omega + q\Phi + \overline{\lambda} \cdot \dot{\overline{a}} + \beta\mathring{s} \tag{6.200}$$

and

$$\nabla\zeta = \nabla\upsilon$$

$$= -\overline{p} + \Sigma_j \lambda_j \nabla a_j + \beta\nabla s \tag{6.201}$$

Because these are exactly the Euler-Lagrange Eqs. 6.170 and 6.171, we have shown that it is possible to derive the Euler-Lagrange equations from the equations of motion. Thus Hamilton's principle does in fact predict the most general kind of motion.

We now have a physical interpretation for the Lagrange multiplier $\overline{\lambda}$. If the flow is irrotational at time t_0, we can take $\overline{\lambda} = 0$; if not we can consider $\overline{\lambda}$ to be the original momentum.

We have just shown that Hamilton's principle leads to Euler-Lagrange equations that are identical in meaning to the ordinary equations of motion. The ordinary equations of motion are nonlinear and contain only physically measurable variables; the Euler-Lagrange equations, on the other hand, are linear (if one considers the quantities that appear as variables) and contain terms with physically unmeasurable potentials and Lagrange multipliers. The Euler-Lagrange equations lead to Manley-Rowe formulas expressed so that the powers can be calculated on the boundary of the system.

Manley-Rowe Formulas. The Manley-Rowe formulas have not previously been given for this system, except for the case of uniform entropy and irrotational flow. The reason the general case could not be done earlier is that the physically unmeasurable quantities $\overline{\lambda}$, $\overline{a}$, and β appear in the formulas; without the motivation of Hamilton's principle we would not be led to define these quantities.

The systematic procedure outlined in Sec. 6.5 yields Manley-Rowe formulas in the form of Eq. 6.2 with the P_α in the form of divergences:

$$P_\alpha = -\nabla \cdot \overline{F}_\alpha \tag{6.202}$$

where

$$\overline{F}_\alpha = \tfrac{1}{2}\mathrm{Re}\Big\{\overline{E}_\alpha \times \overline{H}_\alpha^{\,*} + \iota_\alpha \Big[T_\alpha^* + \Upsilon_\alpha^* + \Big(\frac{\pi}{\rho}\Big)_\alpha^*\Big] + \Sigma_j \Big[\iota_\alpha(\lambda_j \dot{a}_j)_\alpha^* - (\iota\lambda_j)_\alpha \dot{a}_{j\alpha}^*\Big] + \iota_\alpha(\beta\mathring{s})_\alpha^* - (\iota\beta)_\alpha \mathring{s}_\alpha^*\Big\} \tag{6.203}$$

The Manley-Rowe formulas can now be integrated over space, and put into the form of Eq. 6.4. Although the powers can be evaluated at the boundary of the system, the expressions for $\overline{F}_a$ are not what one would expect from a knowledge of the power-flow vector $\overline{F}$ of Eq. 6.159. The physically unmeasurable terms, which disappear for fluid with uniform entropy and irrotational flow, cannot be given a satisfying physical interpretation.

It should be stressed that in deriving the frequency-power formulas, it was assumed that each of the variables could be written in a Fourier series of the form

$$\rho(\overline{r}, t) = \rho_0(\overline{r}) + \text{Re}\, \Sigma_a \rho_a(\overline{r})\, e^{j\omega_a t} \tag{6.204}$$

Whereas there may be physical reasoning to support this assumption for the physically measurable variables, no such reasoning applies to the physically unmeasurable variables, including $\overline{a}$, $\overline{\lambda}$, and (especially) β. It may be true that in most practical cases the flow is such that the formulas hold, but one should make sure of that.

This treatment can be specialized to systems of practical interest simply by omitting the irrelevant terms. For isentropic fluid flow, including acoustics, eliminate the electrical variables by setting $\overline{E}$, $\overline{B}$, $\overline{D}$, $\overline{H}$, q, $\overline{A}$, and Φ all to zero. If the entropy is uniform, drop all terms that involve s, θ, and β. If the flow is irrotational, set $\nabla \times \overline{p}$ to zero, and eliminate all terms involving $\overline{a}$ or $\overline{\lambda}$. On the other hand, if the fluid is incompressible, make ρ a constant; if, in addition, the entropy is uniform, all terms including Υ can be neglected.

For electron clouds, including electron beams, the effects of entropy, gravity, and pressure are usually ignored; therefore, eliminate all terms involving π, s, θ, Υ, β, and Ω. For irrotational flow, ignore the terms containing $\overline{a}$ or $\overline{\lambda}$, and set $\nabla \times \overline{p} = 0$.

6.7 Magnetohydrodynamic Fluid

Here we discuss a magnetohydrodynamic fluid, a fluid with free charges and currents [17]. We assume the conductivity is infinite, so that there are no losses. This assumption forces the electric field in the reference frame moving along with the fluid to vanish. Because this is a significant qualitative difference from the system discussed in Sec. 6.6, one cannot obtain the equations of motion of that system in the limit as the free charges and currents are vanishingly small.

For generality, we include in the analysis all the physical effects discussed in Sec. 6.6, and, in addition, free charges and currents. Thus we allow the fluid to be charged, so that the total electric charge and current is composed of parts caused by the fluid charge and part caused by the free charge. If the fluid charge and current can be neglected compared with the free charge and current, simply put the charge per particle, q, zero. If the entropy is uniform, drop all terms in the analysis that involve the entropy s, the temperature θ, and the Lagrange multiplier β. If the fluid is assumed to be incompressible, simply treat the particle density ρ as a constant; if, in addition, the entropy is uniform, all terms involving the intrinsic energy per particle Υ may be neglected, since Υ is constant.

Instead of repeating much of the description of Sec. 6.6, we merely give the modifications that are necessary for that discussion to apply to systems with free charges and currents and infinite conductivity. Generally speaking, this consists of adding terms to existing equations and adding a few new equations and some new discussion.

Constitutive Relations. In addition to the variables defined in Sec. 6.6 (Constitutive Relations), we have

$$\rho_f = \text{Free charge density} \tag{6.205}$$

$$\overline{J}_f = \text{Free current density} \tag{6.206}$$

Therefore, the total electric charge density is

$$q\rho + \rho_f \tag{6.207}$$

and the total electric current density is

$$q\overline{\iota} + \overline{J}_f \tag{6.208}$$

The constitutive relations discussed in Sec. 6.6 (Constitutive Relations) are unchanged; a new constitutive relation is Ohm's law for the fluid. We assume the fluid is infinitely conducting, so that the electric field within the fluid vanishes. Expressed in terms of the electric and magnetic fields measured in the stationary frame of reference,

$$\overline{E} + \overline{v} \times \overline{B} = 0 \tag{6.209}$$

Equations of Motion. The equations of motion given in Sec. 6.6 must be altered. The forces on the fluid caused by free charge and current modify the force equation, Eq. 6.154, to

$$\nabla\left(T + \Upsilon + \frac{\pi}{\rho} + \Omega\right) = -\dot{\overline{p}} + q\dot{\overline{A}} + q\overline{E} + \overline{v} \times (\nabla \times \overline{p}) + \theta\nabla s + \frac{\rho_f\overline{E}}{\rho} + \frac{\overline{J}_f \times \overline{B}}{\rho} \tag{6.210}$$

The equation of isentropy, Eq. 6.155, and the equation of particle conservation, Eq. 6.156, remain unchanged. The remainder of the equations of motion are Maxwell's equations; Eq. 6.133 and 6.134 are unchanged, but the other two are modified to include the new charge density and current density:

$$\nabla \cdot \overline{D} = q\rho + \rho_f \tag{6.211}$$

$$\nabla \times \overline{H} = q\overline{\iota} + \overline{J}_f + \dot{\overline{D}} \tag{6.212}$$

In addition to these equations of motion is the equation of conservation of free charge,

$$\nabla \cdot \overline{J}_f + \dot{\rho}_f = 0 \tag{6.213}$$

Also, Ohm's law, Eq. 6.209, appears as one of the equations of motion.

The derivation of the power-flow theorem

$$\frac{\partial}{\partial t}(W_e + W_m + \rho T + \rho \Upsilon + \rho\Omega) + \nabla \cdot \overline{F} = 0 \tag{6.214}$$

of Sec. 6.6 remains valid even in the presence of free charges and currents, when Eq. 6.209 is taken into account. The physical power-flow vector $\overline{F}$ is the same as the power-flow vector found in Sec. 6.6, and is given by Eq. 6.159.

<u>The Lagrangian.</u> The Lagrangian of Eq. 6.161 must be modified to include the coupling terms of the free charges and currents. It becomes

$$L = \rho T' - \rho\Upsilon - \rho\Omega + W_e' - W_m - q\rho\Phi + q\overline{\iota} \cdot \overline{A} - \rho_f \Phi + \overline{J}_f \cdot \overline{A} \tag{6.215}$$

Besides the kinematic constraints discussed in Sec. 6.6, we consider Eq. 6.213, the free charge continuity equation, as a kinematic constraint. We choose to take account of it by defining a suitable potential. One possible choice for a potential might be the "pseudo-polarization" vector $\overline{P}_f$ of Panofsky and Phillips [76, Sec. 13-4], with the property that

$$\overline{J}_f = \dot{\overline{P}}_f \tag{6.216}$$

and

$$\rho_f = -\nabla \cdot \overline{P}_f \tag{6.217}$$

We call this a pseudo-polarization vector because of the similarity to the polarization vector $\overline{P}_p$ of a dielectric medium. The polarization charge density ρ_p and the dielectric current density $\overline{J}_p$ are related to the polarization vector $\overline{P}_p$ by [23, Sec. 5.2]

$$\rho_p = -\nabla \cdot \overline{P}_p \tag{6.218}$$

and

$$\overline{J}_p = \dot{\overline{P}}_p \tag{6.219}$$

if the dielectric medium does not move.

However, the use of the vector $\overline{P}_f$ as a potential is not satisfactory when fluid motion is present. We can obtain a hint about the proper potential by noting that in the case of a moving dielectric, the polarization current is no longer given by Eq. 6.219, but rather by [23, Sec. 9.2]

$$\overline{J}_p = \dot{\overline{P}}_p + \nabla \times (\overline{P}_p \times \overline{v}) \tag{6.220}$$

We define a pseudo-polarization vector $\overline{\Pi}$ by analogy with the dielectric case, by requiring that

$$\rho_f = - \nabla \cdot \overline{\Pi} \tag{6.221}$$

and

$$\overline{J}_f = \dot{\overline{\Pi}} + \nabla \times (\overline{\Pi} \times \overline{v}) \tag{6.222}$$

This representation for the free current and charge was suggested by Haus [38].

It is not clear as yet that arbitrary currents can be represented in the form of Eq. 6.222. To realize this, consider the differential equation

$$\frac{D\overline{\Pi}}{Dt} = (\overline{\Pi} \cdot \nabla)\overline{v} - \overline{\Pi}(\nabla \cdot \overline{v}) + \overline{J}_f - \overline{v}\rho_f \tag{6.223}$$

For a given particle, for a given flow, this is a set of three coupled, time-varying, total, differential equations. A unique solution exists with given initial conditions $\overline{\Pi}(t_0)$ specified. Putting together the solution for each particle, we define $\overline{\Pi}(\overline{r}, t)$ as a function of space and time, uniquely, for a given initial condition $\overline{\Pi}(\overline{r}, t_0)$. We select the initial condition in which

$$\nabla \cdot \overline{\Pi}(\overline{r}, t_0) = - \rho_f(\overline{r}, t_0) \tag{6.224}$$

Now Eq. 6.223 can be written in the form

$$\dot{\overline{\Pi}} + \nabla \times (\overline{\Pi} \times \overline{v}) = \overline{J}_f - \overline{v}\rho_f - \overline{v}(\nabla \cdot \overline{\Pi}) \tag{6.225}$$

of which the divergence is

$$\frac{D}{Dt}(\rho_f + \nabla \cdot \overline{\Pi}) = (\rho_f + \nabla \cdot \overline{\Pi})(\nabla \cdot \overline{v}) \tag{6.226}$$

where Eq. 6.213 has been used to eliminate $\overline{J}_f$. But Eq. 6.226 is, for a given particle and given flow, a time-varying, total, differential equation, and has a unique solution. By inspection, a possible solution is

$$\rho_f + \nabla \cdot \overline{\Pi} = 0 \tag{6.227}$$

In view of the initial condition of Eq. 6.224, this is the unique solution. Thus we have defined $\overline{\Pi}(\overline{r}, t)$ in such a way that

$$\rho_f = - \nabla \cdot \overline{\Pi} \tag{6.228}$$

and, from Eq. 6.225,

$$\overline{J}_f = \dot{\overline{\Pi}} + \nabla \times (\overline{\Pi} \times \overline{v}) \tag{6.229}$$

Note that there is no restriction on $\overline{J}_f$ and ρ_f other than the continuity requirement of Eq. 6.213.

When the kinematic constraints are accounted for, the Lagrangian of Eq. 6.215 becomes, instead of Eq. 6.169,

$$\begin{aligned} L = {} & \rho T' - \rho \Upsilon - \rho\Omega + W_e' - W_m - q\rho\Phi + q\overline{\iota} \cdot \overline{A} \\ & + \Phi(\nabla \cdot \overline{\Pi}) + \overline{A} \cdot \dot{\overline{\Pi}} + \overline{A} \cdot \nabla \times (\overline{\Pi} \times \overline{\iota}/\rho) \\ & - \zeta(\dot{\rho} + \nabla \cdot \overline{\iota}) - \beta(\rho\dot{s} + \overline{\iota} \cdot \nabla s) \\ & - \overline{\lambda} \cdot [\rho\dot{\overline{a}} + (\overline{\iota} \cdot \nabla)\overline{a}] \end{aligned} \tag{6.230}$$

which is considered a function of the variables ρ, $\overline{\iota}$, s, $\overline{A}$, Φ, $\overline{a}$, $\overline{\Pi}$, ζ, β, $\overline{\lambda}$, and their space and time derivatives, as well as space explicitly.

Hamilton's Principle. Some of the Euler-Lagrange equations of Sec. 6.6(Hamilton's Principle) are changed, and others are not. Unchanged are Eqs. 6.173, 6.177, 6.178, 6.179, and 6.180. Equations 6.170, 6.171, 6.174, and 6.175 are changed to the formulas:

$$\dot{\zeta} = T + \Upsilon + \frac{\pi}{\rho} + \Omega + q\Phi + \overline{\lambda} \cdot \dot{\overline{a}} + \beta\dot{s} + \frac{\overline{\Pi} \cdot \overline{v} \times \overline{B}}{\rho} \tag{6.231}$$

$$\nabla\zeta = -\overline{p} + \Sigma_j \lambda_j \nabla a_j + \beta\nabla s + \frac{\overline{\Pi} \times \overline{B}}{\rho} \tag{6.232}$$

$$\nabla \times \overline{H} = q\overline{\iota} + \overline{J}_f + \dot{\overline{D}} \tag{6.233}$$

and

$$\nabla \cdot \overline{D} = q\rho + \rho_f \tag{6.234}$$

Also, there is a new Euler-Lagrange equation, resulting from the variation of $\overline{\Pi}$. It is

$$\overline{E} + \overline{v} \times \overline{B} = 0 \tag{6.235}$$

We now have the problem of showing that the Euler-Lagrange equations are equivalent to the ordinary equations of motion. It is easily verified that all the ordinary equations of motion except the force equation follow from the Euler-Lagrange equations. Furthermore, the only Euler-Lagrange equations that require verification are Eqs. 6.231, 6.232, 6.173, 6.177, and 6.180.

It is convenient to introduce a lemma similar to Eq. 6.185. If

$$\overline{J}_f = \dot{\overline{\Pi}} + \nabla \times (\overline{\Pi} \times \overline{v}) \tag{6.236}$$

$$\rho_f = -\nabla \cdot \overline{\Pi} \tag{6.237}$$

$$\nabla \times \overline{E} = -\dot{\overline{B}} \tag{6.238}$$

$$\overline{E} + \overline{v} \times \overline{B} = 0 \tag{6.239}$$

$$\dot{\rho} + \nabla \cdot (\rho \overline{v}) = 0 \tag{6.240}$$

and

$$\nabla \cdot \overline{B} = 0 \tag{6.241}$$

then it is identically true that

$$\frac{\partial}{\partial t}\left(\frac{\overline{\Pi} \times \overline{B}}{\rho}\right) - \nabla \frac{\overline{\Pi} \cdot \overline{v} \times \overline{B}}{\rho} - v \times \left(\nabla \times \frac{\overline{\Pi} \times \overline{B}}{\rho}\right) = \frac{\overline{J}_f \times \overline{B}}{\rho} + \frac{\rho_f \overline{E}}{\rho} \tag{6.242}$$

The force equation can be derived easily now. Take the gradient of Eq. 6.231, and subtract from it the time-derivative of Eq. 6.232, thus eliminating the Lagrange multiplier ζ. Use the two lemmas, Eq. 6.185 and Eq. 6.242, and calculate the quantity

$$\overline{v} \times (\nabla \times \overline{p}) \tag{6.243}$$

from Eq. 6.232, to obtain the force equation in the form of Eq. 6.210. We thus conclude that Hamilton's principle predicts the correct equations of motion.

The generality of Hamilton's principle is shown by predicting the form of the Euler-Lagrange equations from the known equations of motion. We define $\overline{a}(\overline{r}, t)$ and $\beta(\overline{r}, t)$ as in Sec. 6.6. The definition of $\overline{\lambda}(\overline{r}, t)$ is altered only in its initial value; we put $\overline{\lambda}(\overline{r}, t_0)$ equal to any vector so that at time t_0

$$\nabla \times \overline{\lambda} = \nabla \times \left(\overline{p} - \frac{\overline{\Pi} \times \overline{B}}{\rho}\right) \tag{6.244}$$

We have derived all the Euler-Lagrange equations except Eqs. 6.231 and 6.232. Note that the variables $\overline{a}$, $\overline{\lambda}$, s, β, and θ have been chosen so that the lemma of Eq. 6.185 holds, and note that Eqs. 6.236, 6.237, 6.238, 6.239, 6.240, and 6.241 are predicted from the equations of motion and the definition of the pseudo-polarization $\overline{\Pi}$, so that Eq. 6.242 is known to be valid.

Now let us define a vector field $\overline{u}(\overline{r}, t)$ by an extension of the definition of Eq. 6.189:

$$\overline{u} = \nabla \times \left(\overline{p} - \Sigma_j \lambda_j \nabla a_j - \beta \nabla s - \frac{\overline{\Pi} \times \overline{B}}{\rho}\right) \tag{6.245}$$

noting that at time t_0, $\overline{u}$ vanishes everywhere. As redefined here, $\overline{u}$ obeys Eq. 6.192.

The reasoning given in Sec. 6.6 is only slightly altered to fit the new function $\overline{u}(\overline{r}, t)$. We find from the differential equation, Eq. 6.193, that $\overline{u}$ must vanish, so that for some scalar $\upsilon(\overline{r}, t)$,

$$\Sigma_j \lambda_j \nabla a_j + \beta \nabla s + \frac{\overline{\Pi} \times \overline{B}}{\rho} - \overline{p} = \nabla \upsilon \tag{6.246}$$

Using the two lemmas, Eqs. 6.185 and 6.242, and the force equation, Eq. 6.210, this result can be put into the form

$$\nabla\left(T + \Upsilon + \frac{\pi}{\rho} + \Omega + q\Phi\right) = \nabla\dot{\upsilon} - \nabla(\overline{\lambda}\cdot\dot{\overline{a}} + \beta\dot{s}) - \nabla\frac{\overline{\Pi}\cdot\overline{v}\times\overline{B}}{\rho} \tag{6.247}$$

From this point, the same reasoning used in Sec. 6.6, and the same definition for $\zeta(\overline{r}, t)$, lead to the Euler-Lagrange equations, Eqs. 6.231 and 6.232. Thus we have shown that the most general type of motion is predicted by Hamilton's principle.

Manley-Rowe Formulas. The simplest form of the Manley-Rowe formulas is obtained not when L is used as the state function, but rather when

$$L + \nabla\cdot\frac{\overline{A}\times(\overline{\Pi}\times\overline{\iota})}{\rho} \tag{6.248}$$

is used. This differs from L only by a perfect divergence; it is known that both state functions can be used in Hamilton's principle, and lead to the same Euler-Lagrange equations [70, Section 3.4]. The Manley-Rowe formulas generated by this state function are of the form of Eq. 6.4, with the vector $\overline{F}_\alpha$ given by

$$\begin{aligned}\overline{F}_\alpha = \tfrac{1}{2}\mathrm{Re}\Bigg\{&\overline{E}_\alpha\times\overline{H}_\alpha^* + \overline{\iota}_\alpha\left[T_\alpha^* + \Upsilon_\alpha^* + \left(\frac{\pi}{\rho}\right)_\alpha^*\right]\\ &+ \Sigma_j\left[\overline{\iota}_\alpha\left(\lambda_j\dot{a}_j\right)_\alpha^* - \left(\overline{\iota}\lambda_j\right)_\alpha\dot{a}_{j\alpha}^*\right]\\ &+ \overline{\iota}_\alpha(\beta\dot{s})_\alpha^* - (\overline{\iota}\beta)_\alpha\dot{s}_\alpha^*\\ &- \overline{\iota}_\alpha\left(\frac{\overline{\Pi}\cdot\overline{E}}{\rho}\right)_\alpha^* - \overline{E}_\alpha\times(\overline{\Pi}\times\overline{v})_\alpha^*\Bigg\}\end{aligned} \tag{6.249}$$

Observe that this differs from the corresponding result of Sec. 6.6, Eq. 6.203, only by the last two terms. The physical interpretation of these terms is not clear.

PART III

APPLICATIONS

Chapter 7

ROTATING MACHINE APPLICATIONS

It is shown in Sec. 4.6 that some rotating machines (those without commutators) obey the Manley-Rowe formulas if losses within the machine are neglected. Here we demonstrate how these frequency-power formulas can, in simple cases, lead directly to results of practical importance. In Sec. 7.1 we discuss the conventional two-phase induction machine with balanced excitation, and show that the Manley-Rowe formulas predict the existence of three regions of operation — a motor region, a brake region, and a generator region.

In Sec. 7.2 the balanced treatment is extended to cover the unbalanced machine, or the machine run with unbalanced excitation on the stator, but a balanced rotor. There are now four frequencies of interest, so we require an additional constraint; this is supplied by a simple, yet general, rotor model. As special cases of the general solution we discuss the nearly balanced machine and the completely unbalanced machine.

The d-c induction machine is discussed in Sec. 7.3, and the general expressions obtained from the Manley-Rowe formulas are also shown from the actual solution.

7.1 The Balanced Induction Machine

As shown in Sec. 4.6, the induction machine [116, Sec. 3.6.4] obeys the Manley-Rowe formulas, provided stator and rotor loss and mechanical friction are taken out. Of course, rotor loss is essential if the machine is to operate, but for the purpose of applying the Manley-Rowe formulas this resistance must be considered as "outside" the system. This requirement is purely conceptual, however, and the analysis is in no way restricted to machines with a wound rotor, in which this separation can (in part) be carried out physically.

We excite the stator with balanced currents at frequency ω_s. Because the stator and rotor are both assumed balanced, the only current in the rotor windings is at frequency

$$\omega_- = \omega_s - \omega_m \tag{7.1}$$

where ω_m is the (assumed constant) mechanical speed.

If we call P_- the power input to the rotor (that is, the negative of the power dissipated by the rotor resistance), P_s the power input to the stator, and P_m the mechanical power input, and if we consider ω_s and ω_m to be the two independent frequencies, the Manley-Rowe formulas are

$$\frac{P_m}{\omega_m} - \frac{P_-}{\omega_-} = 0 \tag{7.2}$$

and

$$\frac{P_s}{\omega_s} + \frac{P_-}{\omega_-} = 0 \tag{7.3}$$

Without loss of generality, we assume that ω_s is positive; however, either ω_m or ω_- can be negative.

In the normal operation of the induction machine we do not put power into the rotor, so $P_- \leq 0$. For operation as a motor, we put power into the stator and get power out the shaft, so that $P_s > 0$ and $P_m < 0$. For operation as a generator, on the other hand, we put power in the shaft and get power out the stator, so $P_s < 0$ and $P_m > 0$. And finally, there is a third type of operation, known as "brake" operation, in which power flows in both the stator and the shaft, so $P_s > 0$ and $P_m > 0$. These conditions are summarized in Table 7.1.

	P_-	P_s	P_m
MOTOR	Neg.	Pos.	Neg.
GENERATOR	Neg.	Neg.	Pos.
BRAKE	Neg.	Pos.	Pos.

Table 7.1. Balanced machine power flow. Each power shown is the power into the machine

It is clear from Eqs. 7.2 and 7.3 that motor action results only when both ω_m and ω_- are positive, or when

$$0 < \omega_m < \omega_s \tag{7.4}$$

On the other hand, we have generator action only if ω_m is positive and ω_- is negative, that is,

$$\omega_s < \omega_m \tag{7.5}$$

When ω_m is negative we must have a positive ω_-, and Eqs. 7.2 and 7.3 predict that both P_m and P_s are positive, so we get brake operation.

A common dimensionless parameter that expresses the shaft spee

ω_m is the slip s:

$$s = \frac{\omega_-}{\omega_s} \tag{7.6}$$

For negative slip we have a generator, for $0 < s < 1$ we have a motor, and for s greater than 1 a brake is obtained. These results are conveniently summarized in Fig. 7.1, which shows, as a function of ω_m (or of s), the three regions of operation.

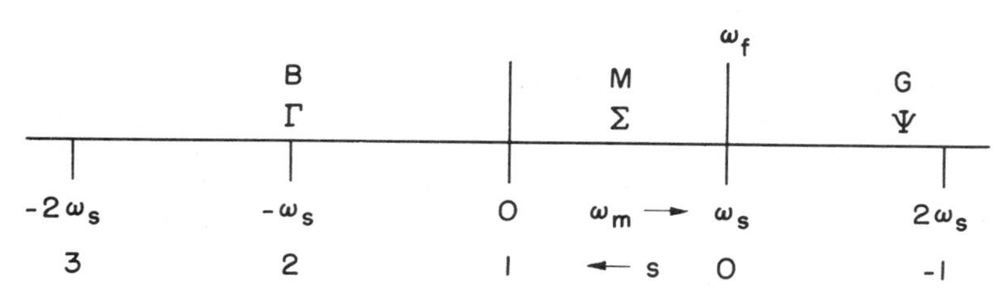

Fig. 7.1. The three regions of operation of a balanced induction machine — the brake region (B), the motor region (M), and the generator region (G). The Greek letters show the correspondence to the more general case of Fig. 7.2

In the motor region, we define the motor efficiency η_m as the ratio of mechanical power out to stator power in, so that by Eqs. 7.2 and 7.3

$$\eta_m = \frac{-P_m}{P_s} = \frac{\omega_m}{\omega_s} = 1 - s \tag{7.7}$$

Furthermore, the starting torque τ_0 is expressed as

$$\tau_0 = \left(\frac{P_m}{\omega_m}\right)_{s\to 1} = -\left(\frac{P_s}{\omega_s}\right)_{s\to 1} \tag{7.8}$$

where the notation $s \to 1$ indicates the limit as $\omega_m \to 0$. A starting torque figure of merit sometimes used is f_0, the negative of the starting torque τ_0 times synchronous speed ω_s, divided by the stator power input P_s required to sustain this torque. In view of Eq. 7.8, this figure of merit for the balanced motor is unity, so for the unbalanced cases discussed in Sec. 7.2, f_0 will express, in a sense, the "starting efficiency" of the motor, as compared to an "equivalent" balanced motor.

The "free-running speed" ω_f is that speed for which the torque vanishes, or the speed the motor approaches if unloaded. This speed is easily found by supposing that the torque is a continuous function of shaft speed. We notice that the applied torque must be negative for motor action and positive for generator action, so the torque must vanish when the slip is zero, or when $\omega_m = \omega_s$. This point is identified on Fig. 7.1 as ω_f.

In the generator region, we define the generator efficiency η_g as the ratio of electrical power out the stator to mechanical power in, so

$$\eta_g = \frac{-P_s}{P_m} = \frac{\omega_s}{\omega_m} = \frac{1}{1 - s} \tag{7.9}$$

Although defined for different regions of operation, the generator efficiency η_g is the reciprocal of the motor efficiency η_m. In terms of η_m, we may classify regions of operation as motor regions $(0 < \eta_m < 1)$, generator regions $(\eta_m > 1)$, and brake regions $(\eta_m <$

These facts about balanced induction machines are well known, and are derived here in this way partly as an indication of the method we use in treating the unbalanced machine in Sec. 7.2.

Use has already been made [86, 85, 21] of the formulas of Eqs. 7.2 and 7.3 for balanced induction generators. References 86, 85, and 21 deal with configurations of coupled balanced induction generators, and show for some particular cases that the efficiency constraints of Eq. 7.9 hold for the overall systems, as well as for each individual machine.

There is some need for a variable-speed, constant-frequency generator for aircraft use, and configurations of induction generators have been proposed in the past to do this at high efficiency, in violation of the Manley-Rowe formulas. Patents have even been issued [86, 85, 21] for devices claiming to do this, although such machines have never been found to work.

It is interesting to note that our general formulation indicates how to get around this efficiency limitation — merely use devices that do not obey the Manley-Rowe formulas. The two simplest such devices are the switch and the rectifier. Efficiency can be improved by using commutator machines, or by rectifying and recovering the otherwise lost rotor power. It cannot be improved, however, by using network of balanced induction machines.

7.2 The Unbalanced Induction Machine

The general analysis of Sec. 7.1 is extended here to unbalanced induction machines [116, Sec. 4.6.3; 25, Art. 10.8]. The same physical device is used for both the balanced and unbalanced induction machines; the difference is in the stator excitation. We no longer require a strictly balanced stator input, although we will insist on a balanced rotor, to limit the number of frequencies present. As in Sec. 7.1, we neglect stator winding resistance, mechanical friction, and other losses, and place the rotor dissipation "outside" the system. Again, however, this requirement is purely theoretical, and the analysis holds for squirrel-cage motors, in which this cannot be done physically, as well as for motors with wound rotors.

In unbalanced operation the shaft turns at a (constant) speed ω_m, and the stator is excited with current of frequency ω_s. The current in the rotor is a sum of currents at frequency

$$\omega_+ = \omega_s + \omega_m \tag{7.10}$$

and at frequency

$$\omega_- = \omega_s - \omega_m \tag{7.11}$$

Let us call P_+ and P_- the power input to the rotor (that is, the negative of the power dissipated by the rotor resistance) caused by currents in the rotor at frequency ω_+ and frequency ω_- respectively. The stator power input, which (because the rotor is assumed to be symmetrical) is entirely at frequency ω_s, is P_s, and the mechanical power input, P_m, is the torque τ multiplied by the speed ω_m.

In an actual machine, the stator power input differs from P_s only because of losses we have neglected, primarily stator resistance; and the mechanical power input differs from P_m only because of mechanical losses.

Because there is power flow only at the four frequencies ω_s, ω_m, ω_+, and ω_-, the frequency-power formulas become

$$\frac{P_m}{\omega_m} + \frac{P_+}{\omega_+} - \frac{P_-}{\omega_-} = 0 \tag{7.12}$$

and

$$\frac{P_s}{\omega_s} + \frac{P_+}{\omega_+} + \frac{P_-}{\omega_-} = 0 \tag{7.13}$$

Just as in the balanced induction machine, no power is put into the rotor. Thus $P_+ \leq 0$ and $P_- \leq 0$. For operation as a motor, we put power in the stator and get power out the shaft; as a generator, power goes in the shaft and out the stator terminals; and as a brake, power flows in both the stator and the shaft.

We put Eqs. 7.12 and 7.13 in the form

$$P_m = \omega_m \left(\frac{P_-}{\omega_-} - \frac{P_+}{\omega_+} \right) \tag{7.14}$$

and

$$P_s = -\omega_s \left(\frac{P_-}{\omega_-} + \frac{P_+}{\omega_+} \right) \tag{7.15}$$

The motor efficiency η_m is defined just as in Sec. 7.1. By the preceeding equations it is

$$\eta_m = \frac{-P_m}{P_s} = \frac{\omega_m}{\omega_s} \frac{P_-\omega_+ - P_+\omega_-}{P_-\omega_+ + P_+\omega_-} \tag{7.16}$$

This efficiency is used to determine the regions of motor action ($0 < \eta_m < 1$), generator action ($\eta_m > 1$), and brake action ($\eta_m < 0$). Table 7.2 summarizes these conditions.

	P_+	P_-	P_s	P_m	η_m
MOTOR	Neg.	Neg.	Pos.	Neg.	$0 < \eta_m < 1$
GENERATOR	Neg.	Neg.	Neg.	Pos.	$\eta_m > 1$
BRAKE	Neg.	Neg.	Pos.	Pos.	$\eta_m < 0$

Table 7.2. Unbalanced induction machine power flow

We notice immediately that if $-\omega_s < \omega_m < \omega_s$, or $0 < s < 2$, then both ω_- and ω_+ are positive, so that the quantity

$$\left| \frac{P_-\omega_+ - P_+\omega_-}{P_-\omega_+ + P_+\omega_-} \right| \tag{7.17}$$

is less than unity. Thus η_m is between -1 and +1, so that the machine is either a motor or a brake. Thus when $-\omega_s < \omega_m < \omega_s$, the machine cannot act as a generator.

Similarly, if ω_m does not lie between $-\omega_s$ and ω_s, either ω_+ or ω_- (but not both) will be negative, so that the quantity of Eq. 7.17 is greater than unity. Thus η_m is either larger than one or less than minus one, so that the machine cannot operate as a motor. It must be either a generator or a brake in these regions.

In words, this means that for shaft speeds less than synchronous speed (in either direction) the device cannot operate as a generator, and for shaft speeds in either direction greater than synchronous speed it cannot act as a motor.

The torque τ is given by

$$\tau = \frac{P_m}{\omega_m} = -\frac{P_s}{\omega_s} \frac{P_-\omega_+ - P_+\omega_-}{P_-\omega_+ + P_+\omega_-} \tag{7.18}$$

so that points of zero torque occur when $P_-\omega_+ = P_+\omega_-$. The starting torque τ_0 is merely the expression of Eq. 7.18 in the limit as the slip $s \rightarrow 1$. The starting torque figure of merit defined in Sec. 7.1 then becomes

$$f_0 = \left(\frac{P_-\omega_+ - P_+\omega_-}{P_-\omega_+ + P_+\omega_-} \right)_{s \rightarrow 1} \tag{7.19}$$

The analysis thus far is general for unbalanced induction machines. To simplify the equations we examine the ratio of $P_-\omega_+$ to $P_+\omega_-$. We adopt a rotor equivalent circuit, which is quite general. In addition, we define an "unbalance parameter" ζ, which describes how close to balanced or unbalanced operation the machine is.

The Rotor Model. Let us break up the magnetic field in the air gap caused by stator currents into a component rotating in the negative direction, and a component rotating in the positive direction. The former generates rotor current at frequency ω_+, and the latter generates rotor current at frequency ω_-. Let ζ be the square of the ratio of the magnitude of these rotating field components. Thus for balanced operation, there is only the field rotating in the positive direction, so $\zeta = 0$. On the other hand, for completely unbalanced operation, that is, either with only one stator winding excited, or with only single-phase current in the stators, we have $\zeta = 1$. For balanced operation in the other direction, $\zeta = +\infty$.

As an example of the calculation of ζ in a practical case, suppose that a two-phase servomotor has one stator excited with current $I_a \cos \omega_s t$ and the other excited with current $I_b \sin \omega_s t$. It is easily seen that

$$\zeta = \left[\frac{I_a - I_b}{I_a + I_b}\right]^2 \tag{7.20}$$

The induced voltage on the rotors is proportional to the time derivative of the field, so using this decomposition of the gap field into rotating components, the ratio (squared) of induced voltages at the two frequencies, E_+ at ω_+, and E_- at ω_-, becomes

$$\left(\frac{E_+}{E_-}\right)^2 = \zeta\left(\frac{\omega_+}{\omega_-}\right)^2 \tag{7.21}$$

As a rotor model we assume an inductance L in series with a resistance R. Since only their ratio is important, they may be defined on a per-phase basis, if desired. The impedances (squared) presented to the rotor current at the sum and difference frequencies are, respectively,

$$Z_+^2 = R^2 + \omega_+^2 L^2 \tag{7.22}$$

and

$$Z_-^2 = R^2 + \omega_-^2 L^2 \tag{7.23}$$

from which we may determine the ratio of rotor currents (squared), and thus the ratio of P_+ to P_-, as

$$\frac{P_+}{P_-} = \frac{E_+^2 Z_-^2 R}{E_-^2 Z_+^2 R} = \zeta\left(\frac{\omega_+}{\omega_-}\right)^2 \frac{R^2 + \omega_-^2 L^2}{R^2 + \omega_+^2 L^2} \tag{7.24}$$

so that

$$\frac{P_+\omega_-}{P_-\omega_+} = \zeta \frac{2 - s}{s} \frac{x^2 + s^2}{x^2 + (2 - s)^2} \tag{7.25}$$

where we define

$$x = \frac{R}{\omega_s L} \tag{7.26}$$

independent of the speed (or slip) of the machine.

Using this rotor model we express the efficiency, torque, etc., in terms of the three parameters x, s, and ζ. These may or may not be independent for any given excitation. The rotor inductance and resistance and the excitation frequency determine x, and so in many cases of practical interest this is a constant. The remaining two parameters, s and ζ, are not, in general, independent. They are (i) for the balanced machine ($\zeta = 0$), (ii) for the completely unbalanced machine ($\zeta = 1$), and (iii) when the stators are driven from current sources. For other types of operation, as for example when one stator winding is excited through a phase-shifting network, or where voltage sources are used, ζ will be found to depend on s. Thus any general examination should take into account the (possibly rather complicated) dependence of ζ on s.

The General Solution. In terms, then, of the parameters x, ζ, and s, we express the motor efficiency as

$$\eta_m = (1 - s) \frac{s[x^2 + (2 - s)^2] - \zeta(2 - s)(x^2 + s^2)}{s[x^2 + (2 - s)^2] + \zeta(2 - s)(x^2 + s^2)} \tag{7.27}$$

the torque as

$$\tau = \frac{-P_s}{\omega_s} \frac{s[x^2 + (2 - s)^2] - \zeta(2 - s)(x^2 + s^2)}{s[x^2 + (2 - s)^2] + \zeta(2 - s)(x^2 + s^2)} \tag{7.28}$$

the starting torque as

$$\tau_0 = \frac{-P_s}{\omega_s} \frac{1 - \zeta}{1 + \zeta} \tag{7.29}$$

and the starting figure of merit as

$$f_0 = \frac{1 - \zeta}{1 + \zeta} \tag{7.30}$$

We find the free-running speed(s) by solving the equation

$$\zeta \frac{2 - s}{s} \frac{x^2 + s^2}{x^2 + (2 - s)^2} = 1 \tag{7.31}$$

and selecting the stable solution(s).

We can interpret these results further by looking for regions of motor, generator, and brake operation, for fixed x and ζ, as a function of s, or shaft speed. We do this under the assumption that ζ is determined independent of s. We further assume, with no loss of generality, that $0 \le \zeta \le 1$. Cases with ζ greater than unity can be obtained by replacing s with $2 - s$ and ζ with $1/\zeta$. Physically this means we consider only cases where the machine "tends" or "prefers" to go in the positive direction; by reversing the definition of mechanical speed we easily treat the other case by symmetry.

For convenience we define the parameter

$$\lambda = \frac{P_+\omega_-}{P_-\omega_+} = \zeta \frac{2 - s}{s} \frac{x^2 + s^2}{x^2 + (2 - s)^2} \tag{7.32}$$

so that

$$\eta_m = (1 - s) \frac{1 - \lambda}{1 + \lambda} \tag{7.33}$$

For MOTOR operation, $0 < \eta_m < 1$. For $0 < s < 1$ this occurs when $-1 < \lambda < 1$, and for $1 < s < 2$ this occurs when $|\lambda| > 1$. For GENERATOR action, $\eta_m > 1$. For $s < 0$ this occurs when $|\lambda| < 1$, and for $s > 2$ this occurs when $|\lambda| > 1$. And finally, for BRAKE operation, $\eta_m < 0$. For $s > 1$ this occurs when $|\lambda| < 1$, and for $s < 1$ this occurs when $|\lambda| > 1$. These conditions are summarized in Table 7.3.

		$\|\lambda\| < 1$	$\|\lambda\| > 1$
$\omega_m < -\omega_s$	$2 < s$	BRAKE	GENERATOR
$-\omega_s < \omega_m < 0$	$1 < s < 2$	BRAKE	MOTOR
$0 < \omega_m < \omega_s$	$0 < s < 1$	MOTOR	BRAKE
$\omega_s < \omega_m$	$s < 0$	GENERATOR	BRAKE

Table 7.3. Unbalanced induction machine operation, in terms of the parameters s and λ

We have reduced the problem to one of determining when λ is less than one in magnitude. A few facts about $\lambda(s)$ bring this out; $\lambda(s)$ has a zero at $s = 2$, and a pole at $s = 0$. Furthermore, $\lambda(s)$ is the ratio of two cubic expressions in s, and so has only five points of zero slope. Since the points of zero slope are also points of zero slope for $1/\lambda(s)$ or $\zeta^2/\lambda(s) = \lambda(2 - s)$, we see that these points are located symmetrically about the point $s = 1$.

For high values of s, λ approaches $-\zeta$ from below, and for high negative values of s, λ approaches $-\zeta$ from above. Because λ is negative except between 0 and 2, there must be exactly one point of zero slope between $-\infty$ and 0, and exactly one point of zero slope between 2 and $+\infty$, for otherwise there would have to be three such points on either side, or a total of six — too many. And between 0 and 2 there may be either two points of zero slope, or no such points.

Because by assumption $\zeta \leq 1$, there must be exactly one (no more no less) point where $\lambda = -1$ in the region $-\infty < s < 0$. Since $\lambda(1) = \zeta \leq 1$, and because there is at most one point of zero slope between 0 and 1, $\lambda = 1$ exactly once in this region. For $1 < s < 2$, λ may equal unity either two times or no times, and for $2 < s < +\infty$, $\lambda = -1$ either two times or no times.

Using this reasoning, we subdivide the range of speeds into regions of operation. The most general type of subdivision of this sort (assuming that ζ is independent of s) is shown in Fig. 7.2. Depending on the values of x and ζ, regions Δ and Ξ may or may not be present. No attempt has been made to have Fig. 7.2, or any other such diagram, drawn to scale.

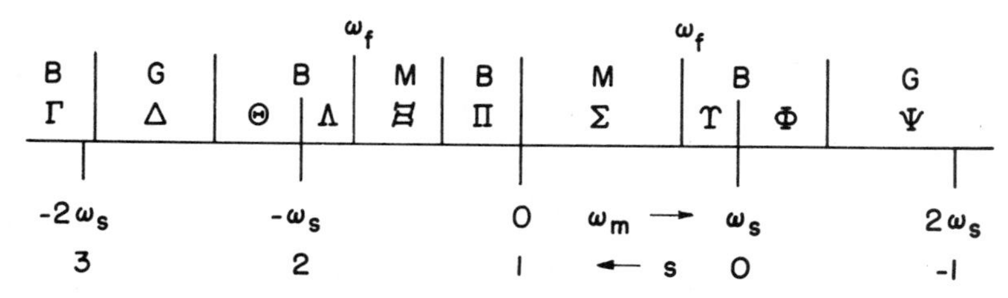

Fig. 7.2. Most general division into regions of operation, under the assumption that x and ζ are independent of the slip

As explained in Sec. 7.2 (Almost-Balanced Operation), the diagram of Fig. 7.2 reduces to that of Fig. 7.1 as balanced operation is approached.

The free-running speeds ω_f are indicated in Fig. 7.2; the point separating regions Ξ and Π is also a point of zero torque, but it is not stable.

Some of the regions of operation shown in Fig. 7.2 are well known. It is not known to what extent, if any, the others, especially the generating and braking regions at high mechanical speed, are important.

Almost-Balanced Operation. This analysis gives a glimpse into the operation of nearly-balanced machines. If a machine is intended to be operated balanced, but there is a defect, either in the device or in the excitation, the operation might be described by a value of

ζ close to, but not quite, zero.

The various regions of operation for the almost-balanced machine are shown in Fig. 7.3. The only difference between Fig. 7.3 and

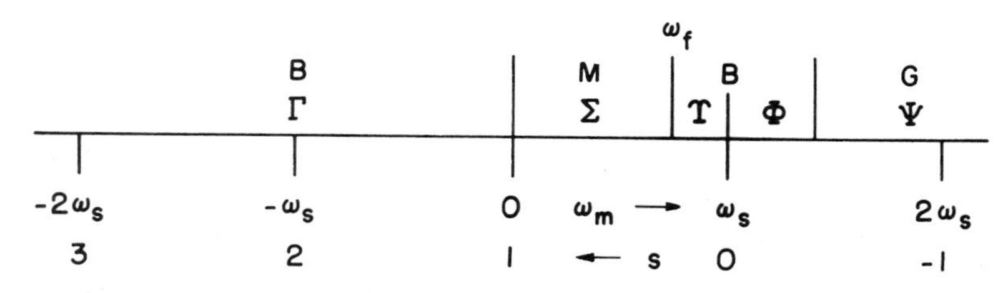

Fig. 7.3. Most general break-up into regions of operation for an almost-balanced machine

Fig. 7.1 (for the balanced machine) is the new braking region close to synchronous speed. The torque, starting torque, starting figure of merit, and efficiency are all continuous functions of ζ, but not uniformly continuous. The almost-balanced machine approximates the balanced machine for ζ small enough, except near regions where the functions have possible singularities. The only such region is close to synchronous speed, and as we have seen, there is an important qualitative difference there. Aside from this, however, the balanced case is obtained in the limit as the unbalance disappears.

In summary, a slight unbalance introduces a small braking region near synchronous speed. The free-running speed is slightly less than synchronous speed. The same effect is obtained when there are small losses in the system, so in practical cases an observed small braking region like this might not be completely eliminated by merely reducing the unbalance.

The Completely-Unbalanced Machine. By "completely-unbalanced" we mean that the air-gap field due to stator currents is a purely pulsating field, so $\zeta = 1$. This occurs when only one stator winding is excited or when the current through all stator windings is of the same phase.

With $\zeta = 1$, the motor efficiency reduces from Eq. 7.27 to

$$\eta_m = (1 - s)^2 \frac{s(2 - s) - x^2}{s(2 - s) + x^2} \tag{7.34}$$

and the torque becomes, from Eq. 7.28

$$\tau = \frac{-P_s(1 - s)}{\omega_s} \frac{s(2 - s) - x^2}{s(2 - s) + x^2} \tag{7.35}$$

The starting torque and starting torque figure of merit both vanish.

The free-running speed is found by setting the torque equal to zero, for nonzero ω_m, that is, by solving the equation

$$s(2 - s) = x^2 \tag{7.36}$$

so

$$\omega_f^{\,2} = \omega_s^{\,2} - \left(\frac{R}{L}\right)^2 \tag{7.37}$$

In this special case of completely unbalanced operation, some of the regions of Fig. 7.2 disappear. The machine is symmetrical in the sense that if it operates in one mode at speed ω_m, the operation at speed $-\omega_m$ will be identical; there is no "preferred" direction of rotation. As a consequence, regions Γ and Π must vanish; the most general chart of the form of Fig. 7.2 for the completely unbalanced machine is Fig. 7.4. The regions Ξ and Σ vanish if $x > 1$, or if $R > \omega_s L$. Note the free-running speeds ω_f are indicated.

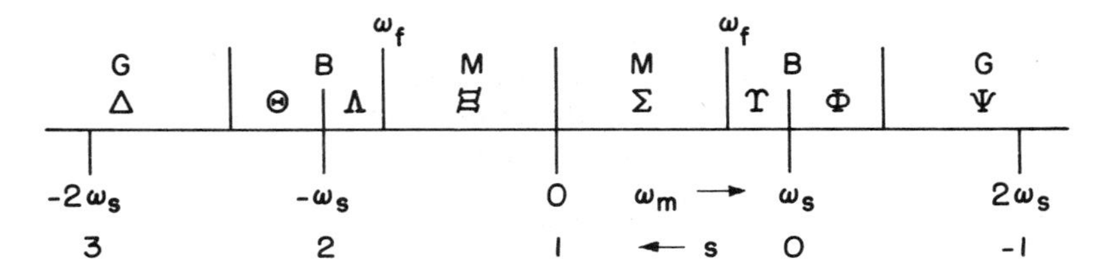

Fig. 7.4. Most general division into regions of operation for a completely unbalanced machine. The free-running speeds ω_f are indicated. Note the symmetry about the point $s = 1$

7.3 The d-c Induction Machine

The d-c induction machine [116, Sec. 4.4.2] uses a commutator in an essential way, but nevertheless we can derive an expression for efficiency of the device from the Manley-Rowe formulas. We must be careful to define our system in such a way that the commutator and brushes are outside. The commutator is used to convert frequency; hence the frequencies applied to the system we are allowed to consider, do not coincide with the frequencies applied to the actual machine. Direct current is applied to the rotor brushes in the actual operation, but we have to consider the power input at these terminals to be a-c.

In the operation of the device, the brush carriage, with position indicated by $\psi(t)$, is rotated at a constant rate, so that

$$\psi(t) = \omega_b t \tag{7.38}$$

and the rotor, whose position is $\phi(t)$, also rotates at a constant speed ω_m. If a d-c current I is injected into the γ rotor brushes and the δ rotor brushes are left open-circuited, there will be a net torque developed and the device will operate as a motor, provided the stator windings are terminated in resistors.

We now inquire as to the power relations. The current and voltages in the stator will be of frequency the same as that supplied by the rotating field, ω_b. Furthermore, the motor driving the brush carriage does no work on the system other than to overcome frictional losses, which in this analysis we neglect anyway, so the power input from that source can be neglected.

The current through a given rotor wire will not be direct current, in spite of the fact that the rotor is supplied with d-c. The relative speed of the commutator brushes over the rotor is merely $\omega_b - \omega_m$, and the current each rotor winding experiences is a square wave with this fundamental frequency. Thus for application of the Manley-Rowe formulas to the d-c induction motor, we must consider the rotor power to be at frequency $\omega_b - \omega_m$ and all its odd harmonics.

However, in applying the formulas it is known that we may consider power at harmonics of a given frequency to be power at that given frequency (see Appendix C). Thus we consider all the rotor power input to be at the fundamental frequency $\omega_b - \omega_m$, and not spread among the various harmonics, as it physically is.

Our procedure is then clear. Power to the rotor we consider to be at frequency $\omega_r = \omega_b - \omega_m$; power to the stator is at frequency ω_b; and power at the shaft is d-c power, which by Appendix D we may consider to be power at frequency ω_m.

Let the rotor, stator, and shaft power inputs be, respectively, P_r, P_s, and P_m. Then the Manley-Rowe formulas are

$$\frac{P_s}{\omega_b} + \frac{P_m}{\omega_m} = 0 \tag{7.39}$$

and

$$\frac{P_r}{\omega_r} - \frac{P_m}{\omega_m} = 0 \tag{7.40}$$

For motor operation we want power to flow in only at the rotor, so that $P_r > 0$, $P_m < 0$, and $P_s < 0$. A glance at Eqs. 7.39 and 7.40 shows that this is possible, for $\omega_m < 0$; in fact, the motor efficiency η_m is

$$\eta_m = -\frac{P_m}{P_r} = \frac{(-\omega_m)}{\omega_b + (-\omega_m)} \tag{7.41}$$

which shows that the faster the rotor turns over, the higher the efficiency. Equation 7.41 further shows that for $0 < \omega_m < \omega_b$ the device runs as a brake, and for $\omega_b < \omega_m$ it acts as a generator.

The predicted power relationships are verified by the exact solution [116, Sec. 4.4.2]. The power output $(-P_m)$ is given by $T_{e\mu}\omega_m$, where $T_{e\mu}$ is the mechanical torque of electrical origin. The stator power output $(-P_s)$ is given by $2(\frac{1}{2})R^s|i^s_\alpha|^2$, the factor of 2 arising from the two stator windings with identical power, and the factor of one-half coming from averaging the sinusoidal waveform of i^s_α and i^s_β. Using the formulas for stator current and torque given in reference 116, Eqs. 4.170, 4.171, and 4.172, we can immediately verify Eq. 7.39. Equation 7.40 can be verified from the expression for rotor input power

$$P_r = v^r_\gamma i^r_\gamma = v^r_\gamma I \tag{7.42}$$

The brush voltage v^r_γ is not given explicitly in reference 116, but can be calculated easily from Eq. 4.136a of reference 116 and the solutions for the stator currents, Eqs. 4.170 and 4.171. It turns out to be (neglecting rotor resistance)

$$v^r_\gamma = \frac{\left(L^{sr}_\mu\right)^2 n^2 \omega_b\left(\omega_b - \omega_m\right) I R^s}{\left(R^s\right)^2 + \left(n\omega_b L^s_\mu\right)^2} \tag{7.43}$$

from which Eq. 7.40 is immediately verified.

Chapter 8

COMMUNICATIONS APPLICATIONS

Frequency-power formulas are conservation theorems, and are of the most value when used as such: to find constants of the motion, to base new definitions on, to check analytic results, and, especially, to give an intuitive grasp of the frequency conversion operation. Only occasionally can the formulas be used to obtain directly results of practical interest, and these cases usually involve only a few frequencies. A few typical applications to communications systems are discussed here. We cannot, of course, pretend to discuss all applications or even to discuss any thoroughly.

In Sec. 8.1 the constraints imposed by the formulas on harmonic generators are given; it is found from the real-power constraints that nonlinear reactance generators are not limited in efficiency by the frequency-power formulas, whereas nonlinear resistor generators are. These results are not new, but the reactive-power constraints given have not, to our knowledge, been used.

Many frequency converters are used in communications for processing small signals, and linearized time-varying equations of motion are almost universally used in analyzing such applications. In Sec. 8.2 we show how these linear time-varying equations are derived for a parametrically-pumped nonlinear system, of the type that obeys any of the four types of frequency-power formulas; we also give the resulting frequency-power formulas. Many such devices are analyzed by matrix methods, and in Sec. 8.3 the conversion matrices are shown to be constrained by the frequency-power formulas. In Sec. 8.4. we demonstrate a typical use of frequency-power formulas to determine fundamental limits of performance of a special type of frequency converter. We do not treat even this small problem completely, but merely show one method by which the frequency-power formulas can be used.

Several other interesting applications of the Manley-Rowe formulas to communications have been made already. We will not discuss these in detail, but only briefly mention them here.

If there are two independent frequencies and only three frequencies at which power is exchanged, then the frequency-power formulas are very simple, and in fact ratios of the powers can be calculated. Rowe [60, 89], Uhlir [107], Duinker [18], and others have discussed, from this point of view, the use of nonlinear reactances as parametric amplifiers, upconverters, modulators, mixers, downconverters, etc.

Uhlir [107] has made an interesting observation about the use of the Manley-Rowe formulas in three-frequency devices. He compares them to other "second laws," such as the second law of thermodynamics, or the law of conservation of momentum in classical particle dynamics. These "second laws," together with the law of conservation of energy, lead directly to expressions of interest, such as efficiencies, in the same way the Manley-Rowe formulas do. For example, one can combine the first and second laws of thermodynamics into a form similar to Eq. 1.12.

$$\frac{\text{Heat Input}}{T_h} = \frac{\text{Work Output}}{T_h - T_c} = \frac{\text{Heat Rejected}}{T_c} \tag{8.1}$$

for a reversible Carnot engine, where T_h and T_c are the temperatures of the hot and cold resevoirs.

The theory of noise in parametric amplifiers is both simplified and made more general by using the Manley-Rowe formulas [40, 93]. The theory is simplified because the formulas can be used in place of detailed solutions, and it is made more general because specific circuits need not be used, and because the results are shown to hold for any physical system that satisfies the Manley-Rowe formulas.

Kurokawa and Hamasaki [50, 49] have made interesting use of the formulas by showing that they predict a type of orthogonality of solutions of a parametrically-pumped distributed system, in a way similar to the way the ordinary Poynting's theorem predicts orthogonality of modes in lossless distributed systems [4]. The nonlinear theory of a limiter made from a weakly-pumped reactance was discussed by Olson, Wang, and Wade [72], who used the Manley-Rowe formulas to simplify their derivation.

One of the most interesting applications is to magnetic amplifiers. A magnetic amplifier can be thought of as a combination of a nonlinear reactance upconverter and a nonlinear resistance downconverter, and Manley [58] has published a discussion along these lines, using his formulas.

8.1 Harmonic Generators

Nonlinear elements can be used as harmonic generators, and it is interesting to ask what constraints, if any, the frequency-power formulas impose. Power is put into the device only at frequency ω, and we assume there is no other independent frequency present (note that this last statement is an assumption, not a statement of fact). Power then flows out only at frequencies of the form $k\omega$, with k positive. If P_k and Q_k are the real and reactive powers into the device at frequency $k\omega$, the four frequency-power formulas reduce to

$$\text{Type I:} \qquad \Sigma_k P_k = 0 \tag{8.2}$$

$$\text{Type II:} \qquad \Sigma_k k Q_k = 0 \tag{8.3}$$

Type III (Page form): $\Sigma_k P_k(1 - \cos k\theta) \geq 0$ (8.4)

Type III (Pantell form): $\Sigma_k k^2 P_k \geq 0$ (8.5)

Type IV (Inductive): $\Sigma_k Q_k(1 - \cos k\theta)/k \geq 0$ (8.6)

$\Sigma_k kQ_k \geq 0$ (8.7)

Type IV (Capacitive): $\Sigma_k Q_k(1 - \cos k\theta)/k \leq 0$ (8.8)

$\Sigma_k kQ_k \leq 0$ (8.9)

Equation 8.2 is merely the law of conservation of energy, and the Manley-Rowe formulas have nothing more than this to say. Because of this, it is often thought that efficiency arbitrarily close to 100% is possible for a nonlinear capacitor harmonic generator[13, 52, 53]. Actually, such a conclusion must [46] come from a circuit analysis; the Manley-Rowe formulas do not say that this is possible — they merely do not say that it is impossible.

On the other hand, for nonlinear resistance harmonic generators, the frequency-power formulas of the third type, Eq. 8.5, give a definite restriction on harmonic generation efficiency. The power out $-P_n$ at frequency $n\omega$ is bounded by Eq. 8.5:

$$-P_n \leq \frac{\Sigma_k k^2 P_k}{n^2} \leq \frac{P_1}{n^2} \tag{8.10}$$

where the sum is over all k except n; thus the conversion efficiency is limited to $(1/n^2)$, as reported specifically by Page[74,75] and noted by others.

As far as we know, the second and fourth types of frequency-power formulas have never been applied to harmonic generators.

Subharmonics can be generated with nonlinear reactances[52, 53]; if power is put in at frequency ω, power can be extracted at frequencies $\omega/2$, $\omega/3$, etc. Note that the Manley-Rowe formulas, Eq. 8.2, do not prevent this. Much of the parametric excitation observed in the 19th century is in reality subharmonic generation. The parametron computing element [109, 31, 117] is one practical device that makes use of this phenomenon.

Subharmonic generation is impossible, however, with nonlinear resistors [74] that obey frequency-power formulas of Type III. Pantell's form does not predict that it is impossible, but Page's form does. Setting $\theta = 2\pi$ in Eq. 8.4, we find a weighted sum of power inputs at all frequencies except ω and its harmonics, is positive. Because by assumption power is put in only at frequency ω, it can therefore come out only at harmonics of ω. Moreover, as Page has shown [74], if power is put in a device that satisfies frequency-power formulas of Type III only at the two frequencies ω_1 and ω_0, power can come out only at frequencies of the form $m\omega_1 + n\omega_0$, with m and n integers.

8.2 Linear Parametrically-Pumped System

A nonlinear system of the type discussed in this report, when strongly pumped, appears, to superimposed small signals, just like a linear time-varying system. In fact, one can even calculate the parameters of the time-varying system, from a knowledge of the actual nonlinear system and the pumping. As an example of this, we consider a case of some practical interest, a nonlinear capacitor pumped by a voltage $e_p(t)$, given as a function of time. The treatment can then be generalized. We will see that if the overall nonlinear system can be shown to obey any of the four types of frequency-power formulas by the proofs of Chapter 3, then the equivalent time-varying system obeys formulas of the same type.

The nonlinear capacitor is characterized by its charge q being a function of the voltage e; let us define the "incremental capacitance"

$$C_i(e) = \frac{dq}{de} \tag{8.11}$$

as a function of e. Now let the unperturbed variables (voltage, charge, and current), as pumped, have a subscript p (for pump), and the small perturbations have a subscript s (for small-signal); thus

$$e(t) = e_p(t) + e_s(t) \tag{8.12}$$

$$q(t) = q_p(t) + q_s(t) \tag{8.13}$$

and

$$i(t) = i_p(t) + i_s(t) \tag{8.14}$$

where

$$i_s(t) = \frac{dq_s(t)}{dt} \tag{8.15}$$

Our goal is to obtain equations of motion involving the small-signal variables alone.

This is easily done; we expand the charge in a Taylor's series about the unperturbed solution:

$$q(t) = q_p(t) + \left(\frac{dq(e)}{de}\right)_p e_s(t) + \ldots \tag{8.16}$$

where the subscript p on the derivative denotes evaluation at the unperturbed solution. We note from Eqs. 8.13 and 8.15, then, that

$$i_s(t) = \frac{d}{dt}\left[\left(\frac{dq(e)}{de}\right)_p e_s(t)\right] \tag{8.17}$$

if the small signals are small enough, which is in the desired form

$$i_s(t) = \frac{d}{dt}\left[C(t)\, e_s(t)\right] \tag{8.18}$$

if we identify the equivalent time-varying capacitance with the incremental capacitance, as pumped,

$$C(t) = C_i\left(e_p(t)\right) \tag{8.19}$$

We have succeeded, then, in replacing a heavily-pumped nonlinear system with a linear time-varying system that is equivalent, as far as small perturbations are concerned. It is clear that we cannot, within the framework of the time-varying model alone, predict its range of validity, nor can we predict any of the very interesting limiting and saturating properties the finitely-pumped nonlinear system would exhibit.

This technique can also be used with more complicated systems. It is our intent to show that it can be used for any system for which any of four proofs in Chapter 3 apply. Although we use notation appropriate to systems that obey the first type of frequency-power formula, the reader can easily check that the following paragraph also applies to systems that can be discussed in Secs. 3.3, 3.4, or 3.5.

Suppose we have a number of variables x_i, on which an equal number of variables f_j depend. Then, suppose the system described by these variables is pumped with the variables $x_{ip}(t)$ and $f_{jp}(t)$. We wish to replace the heavily-pumped nonlinear system with a linear time-varying system that is equivalent as far as the small-signal variables $x_{is}(t)$ and $f_{js}(t)$ are concerned. This is easily done by writing a Taylor's series expansion for $f_j(t) = f_{jp}(t) + f_{js}(t)$:

$$f_j(t) = f_{jp}(t) + \Sigma_i\left(\frac{\partial f_j}{\partial x_i}\right)_p x_{is}(t) + \ldots \tag{8.20}$$

Taking only the first-order terms, we find linear time-varying equations of motion,

$$f_{js}(t) = \Sigma_i\left(\frac{\partial f_j}{\partial x_i}\right)_p x_{is}(t) \tag{8.21}$$

which accurately describe the small perturbations of the system if they are small enough, and if the higher derivatives of the f_j with respect to the x_i are small enough. Note in particular that the perturbations must be small for all time, so that if the $x_i(t)$ are of the form of Eq. 3.17, then v_{i0}, the average value of $dx_i(t)/dt$, must not change, for otherwise x_{is} would not be bounded.

We could have allowed significant space variation of the variables, so this derivation is not limited to lumped systems. In the description of many systems in Chapter 6, however, physically-unmeasurable potentials and Lagrange multipliers are used, and it is not certain that these variables can be decomposed into an unperturbed par and a bounded perturbation, even if the measurable variables can be One must therefore use caution in dealing with pumped distributed systems.

We now wish to show that the time-varying model thus produced obeys the same type of frequency-power formulas that the pumped nonlinear system does; or, more precisely, that if any of the four proofs of Chapter 3 holds for the nonlinear system, then the same proof holds for the equivalent time-varying system, with the expressions for real or reactive power being of the same form. We do this for each of the four types in turn.

First, suppose the system has an energy state function $U(x_i)$, with

$$f_j = \frac{\partial U(x_i)}{\partial x_j} \tag{8.22}$$

so that the proof of the Manley-Rowe formulas of Sec. 3.2 holds. In that case,

$$\left(\frac{\partial f_j}{\partial x_i}\right)_p = \left(\frac{\partial f_i}{\partial x_j}\right)_p \tag{8.23}$$

and we may evaluate the line integral

$$\Sigma_i \int f_{is} dx_{is} \tag{8.24}$$

using Eq. 8.21, by any path and get the same result,

$$\begin{aligned} U_t &= \tfrac{1}{2}\Sigma_i f_{is} x_{is} \\ &= \tfrac{1}{2}\Sigma_i \Sigma_j \left(\frac{\partial f_i}{\partial x_j}\right)_p x_{is} x_{js} \end{aligned} \tag{8.25}$$

Thus U_t is a state function of the x_{is} and time explicitly (through the pumping), with

$$f_{js} = \frac{\partial U_t(x_{is}, t)}{\partial x_{js}} \tag{8.26}$$

so the time-varying small-signal system must obey the first type of frequency-power formula, with the P_α given by the same form, Eq 3.25, but with subscripts s on the variables:

$$P_\alpha = \tfrac{1}{2}\text{Re}\ \Sigma_i\ j\omega_\alpha (f_{is})_\alpha^* (x_{is})_\alpha \tag{8.27}$$

It does not, of course, obey any such formula corresponding to an independent frequency on which the pump waveform depends. Note that there is no d-c contribution to the time-varying system's formulas.

When these steps are applied to a distributed system that obeys Hamilton's principle, it is found that the resulting P_α can always be expressed as the divergence of a vector of the form of Eq. 6.132, but involving only the small-signal components of the variables:

$$P_\alpha = -\nabla \cdot \overline{F}_\alpha \tag{8.28}$$

where each component of $\overline{F}_\alpha$ is

$$F_{\alpha j} = \tfrac{1}{2}\mathrm{Re}\,\Sigma_i \left(\frac{\Delta L}{\Delta \dfrac{\partial \eta_i}{\partial x_j}} \right)^*_{s\alpha} (\dot{\eta}_{is})_\alpha \tag{8.29}$$

The same sort of analysis holds for systems known to obey the second type of frequency-power formula. Here, the dissipation function is $G(v_i)$, with

$$f_j = \frac{\partial G(v_i)}{\partial v_j} \tag{8.30}$$

and therefore,

$$\left(\frac{\partial f_i}{\partial v_j} \right)_p = \left(\frac{\partial f_j}{\partial v_i} \right)_p \tag{8.31}$$

Thus the dissipation function

$$\begin{aligned} G_t &= \tfrac{1}{2}\,\Sigma_i f_{is} v_{is} \\ &= \tfrac{1}{2}\,\Sigma_i \Sigma_j \left(\frac{\partial f_i}{\partial v_j} \right)_p v_{is} v_{js} \end{aligned} \tag{8.32}$$

is a state function of the v_{is} and time explicitly, with

$$\frac{\partial G_t(v_{is}, t)}{\partial v_{js}} = f_{js} \tag{8.33}$$

and so the linear time-varying system obeys the second type of frequency-power formula, with the Q_α of the form of Eq. 3.26 with s subscripts:

$$Q_\alpha = \tfrac{1}{2}\,\mathrm{Im}\,\Sigma_i (f_{is})_\alpha^* (v_{is})_\alpha \tag{8.34}$$

Also, if the original nonlinear system obeys the third type of frequency-power formula, the time-varying model for the pumped non-linear system also obeys this type, provided the pump does not force the nonlinear system into regions in which the matrix of Eq. 3.51 is not positive definite. The expression for the small-signal variables

$$f_{js} = \Sigma_i \left(\frac{\partial f_j}{\partial v_i} \right)_p v_{is} \tag{8.35}$$

gives each f_{js} as a function of the v_{is} and time explicitly (through the known functions $(\partial f_j / \partial v_i)_p$). It is then known from the proof of Sec. 3.4 that frequency-power formulas of the third type hold, provided the matrix with entries

$$\frac{\partial f_{js}(v_{is}, t)}{\partial v_{is}} = \left(\frac{\partial f_j}{\partial v_i} \right)_p \tag{8.36}$$

is positive definite, which it is if the pump does not force the original nonlinear system into regions where the matrix of Eq. 3.51 is not positive definite.

And finally, a system that is shown to obey the frequency-power formulas of the fourth type by the proof of Sec. 3.5, if pumped heavily, can be replaced for small signals by a linear time-varying model; and this model itself obeys the same type of formulas provided the pump does not force the original nonlinear system into regions where the matrix of Eq. 3.66 is not positive definite.

These results can be summarized as follows:

> Let a system that can be shown by the methods of Chapter 3 to obey one type of frequency-power formulas be pumped strongly. Then, to small signals, the pumped nonlinear system behaves like a linear time-varying system, whose parameters can be calculated from a knowledge of the nonlinear system and the pumping. Also, the equivalent linear time-varying system obeys frequency-power formulas of the same type, where the expression for real or reactive power is of the same form, but is calculated from the first-order small-signal variables alone.

With this theorem established, we can write explicitly the four types of frequency-power formulas for linear, parametrically-pumped systems. Let us suppose that pumping is at frequency ω_p and its harmonics, and that signal and generated sidebands appear only at frequencies of the form $k\omega_p + \omega_s$, where k is positive or negative. Denoting the real and reactive power at frequency $k\omega_p + \omega_s$ by P_k and Q_k respectively, we have

Type I:
$$\Sigma_k \frac{P_k}{k\omega_p + \omega_s} = 0 \tag{8.37}$$

Type II: $$\Sigma_k Q_k = 0 \tag{8.38}$$

Type III: $$\Sigma_k P_k \geq 0 \tag{8.39}$$

Type IV (Inductive): $$\Sigma_k \frac{Q_k}{k\omega_p + \omega_s} \geq 0 \tag{8.40}$$

Type IV (Capacitive): $$\Sigma_k \frac{Q_k}{k\omega_p + \omega_s} \leq 0 \tag{8.41}$$

Note that in Eq. 8.37 there is no contribution from d-c power, because the small-signal components of the variables necessary to generate it must vanish.

If k is far enough negative so that $k\omega_p + \omega_s$ is negative, then we may wish to sum over positive frequencies. This may be done in Eqs. 8.37, 8.39, 8.40, and 8.41, because the real power associated with a negative frequency is the same as the real power associated with the same positive frequency. However, Eq. 8.38 must be replaced by

$$\Sigma_k \frac{k\omega_p + \omega_s}{|k\omega_p + \omega_s|} Q_k = 0 \tag{8.42}$$

if we are to sum over the corresponding positive frequencies, because the reactive power associated with a negative frequency is the negative of the reactive power associated with the corresponding positive frequency. In Eq. 8.42 the Q_k are to be interpreted as reactive power inputs at frequency $|k\omega_p + \omega_s|$.

8.3 Matrix Constraints

Much linear time-invariant single-frequency circuit analysis is done by matrix methods. For example, to describe an n-port linear time-invariant network such as the one in Fig. 8.1, we may use an $n \times n$ impedance matrix, or admittance matrix, or scattering matrix, or other type of matrix. The same techniques are useful in describing linear parametrically-pumped networks.

Just as conditions of losslessness, passivity, etc., place constraints on the matrices that describe linear time-invariant networks, so the appropriate frequency-power formulas place constraints on the matrices that describe time-varying networks of certain types. In Sec. 8.3 (Time-Invariant Matrix Constraints), we derive the well known matrix constraints for linear time-invariant networks of various types, as an indication of the method used. Then in Sec. 8.3 (Linear Time-Varying Matrix Constraints), we derive the constraints on matrices for time-varying systems of the various types. Some of these constraints can be expressed in a variety of forms, which we tabulate for convenience. A few of these time-varying system matrix constraints have been used in the past for special purposes [40, 93, 50, 49].

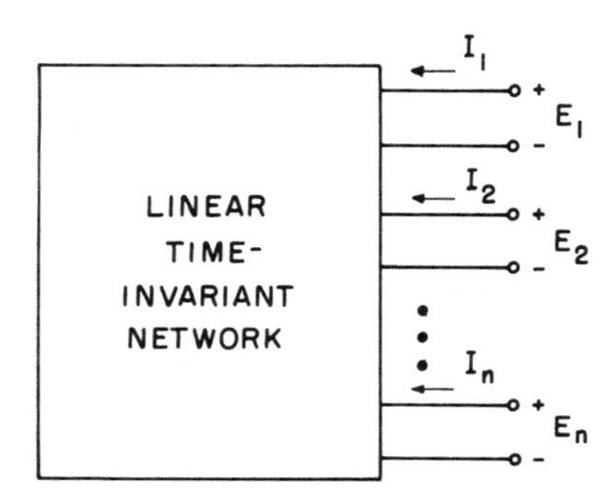

Fig. 8.1. Schematic representation of a linear time-invariant system, with n ports. Although electrical variables are shown, some of the ports may be mechanical

Time-Invariant Matrix Constraints. Consider the network of Fig. 8.1. It has n ports, each of which has a voltage of frequency ω and (complex) amplitude E_i ($i = 1, 2, \ldots n$), and current of amplitude I_i. If we define the column matrix $\underset{\sim}{E}$ with entries E_i, in order, and the column matrix $\underset{\sim}{I}$ with entries I_i, in order, we may express the linear equations of motion of the terminal variables in the form

$$\underset{\sim}{E} = \underset{\sim}{Z}\underset{\sim}{I} \tag{8.43}$$

where the $n \times n$ square matrix $\underset{\sim}{Z}$ is the impedance matrix of the network. The admittance matrix is $\underset{\sim}{Y} = \underset{\sim}{Z}^{-1}$

$$\underset{\sim}{I} = \underset{\sim}{Y}\underset{\sim}{E} \tag{8.44}$$

Alternatively, we may define the scattering variables for each port

$$a_i = \frac{V_i + Z_i I_i}{2\sqrt{Z_i + Z_i^*}} \tag{8.45}$$

and

$$b_i = \frac{V_i - Z_i^* I_i}{2\sqrt{Z_i + Z_i^*}} \tag{8.46}$$

where Z_i is an arbitrary complex number with positive real part (often, for convenience, Z_i is taken to be the terminating impedance at port i). The linear equations of motion can then be written in the form

$$\underset{\sim}{b} = \underset{\sim}{S}\underset{\sim}{a} \tag{8.47}$$

where $\underset{\sim}{b}$ is the column matrix of b_i, $\underset{\sim}{a}$ is the column matrix of a_i, and the $n \times n$ square matrix $\underset{\sim}{S}$ is called the scattering matrix [11].

The total power input P is the sum of the power inputs at each port, or

$$\begin{aligned} P &= \tfrac{1}{2}\mathrm{Re}\, \Sigma_i\, E_i I_i^* \\ &= \tfrac{1}{4}\Sigma_i (E_i I_i^* + E_i^* I_i) \end{aligned} \tag{8.48}$$

which can be written in matrix form as

$$\begin{aligned} P &= \tfrac{1}{4}(\underset{\sim}{E}\dagger \underset{\sim}{I} + \underset{\sim}{I}\dagger \underset{\sim}{E}) \\ &= \tfrac{1}{4}\underset{\sim}{E}\dagger (\underset{\sim}{Y}\dagger + \underset{\sim}{Y})\underset{\sim}{E} = \tfrac{1}{4}\underset{\sim}{I}\dagger (\underset{\sim}{Z}\dagger + \underset{\sim}{Z})\underset{\sim}{I} \end{aligned} \tag{8.49}$$

where the dagger indicates the complex conjugate transpose of a matrix; or P can be written in terms of the scattering variables as

$$\begin{aligned} P &= \Sigma_i(|a_i|^2 - |b_i|^2) \\ &= \underset{\sim}{a}\dagger\underset{\sim}{a} - \underset{\sim}{b}\dagger\underset{\sim}{b} \\ &= \underset{\sim}{a}\dagger(\underset{\sim}{1} - \underset{\sim}{S}\dagger\underset{\sim}{S})\underset{\sim}{a} \end{aligned} \tag{8.50}$$

where $\underset{\sim}{1}$ is the unit matrix. Similarly, the reactive power input Q can be written

$$\begin{aligned} Q &= \frac{1}{4j}\Sigma_i(E_i I_i^* - I_i E_i^*) = \frac{1}{4j}(\underset{\sim}{I}\dagger\underset{\sim}{E} - \underset{\sim}{E}\dagger\underset{\sim}{I}) \\ &= \frac{1}{4j}\underset{\sim}{E}\dagger(\underset{\sim}{Y}\dagger - \underset{\sim}{Y})\underset{\sim}{E} = \frac{1}{4j}\underset{\sim}{I}\dagger(\underset{\sim}{Z} - \underset{\sim}{Z}\dagger)\underset{\sim}{I} \end{aligned} \tag{8.51}$$

The properties of the network place constraints on the matrices $\underset{\sim}{Z}$, $\underset{\sim}{Y}$, and $\underset{\sim}{S}$. First, suppose the network is lossless. Then P = 0, for any choice of $\underset{\sim}{E}$, $\underset{\sim}{I}$, or $\underset{\sim}{a}$, so that from Eq. 8.49

$$\underset{\sim}{Z} + \underset{\sim}{Z}\dagger = 0 \tag{8.52}$$

and

$$\underset{\sim}{Y} + \underset{\sim}{Y}\dagger = 0 \tag{8.53}$$

and from Eq. 8.50

$$\underset{\sim}{1} - \underset{\sim}{S}\dagger\underset{\sim}{S} = 0 \tag{8.54}$$

or $\underset{\sim}{S}$ is unitary [43, Sec. 1.16]. Note that Eqs. 8.52, 8.53, and 8.54

do not require that $\underset{\sim}{Z}$, $\underset{\sim}{Y}$, or $\underset{\sim}{S}$ be symmetrical.

Secondly, suppose that the network is pure lossy, in which case the total reactive power input Q vanishes. Then, from Eq. 8.51,

$$\underset{\sim}{Z} = \underset{\sim}{Z}^{\dagger} \tag{8.55}$$

and

$$\underset{\sim}{Y} = \underset{\sim}{Y}^{\dagger} \tag{8.56}$$

or both $\underset{\sim}{Z}$ and $\underset{\sim}{Y}$ are Hermitian [43, Sec. 1.16].

As a third example, suppose that the network is passive. Then P is nonnegative, so that from Eq. 8.49

$$\underset{\sim}{Z} + \underset{\sim}{Z}^{\dagger} \geq 0 \tag{8.57}$$

and

$$\underset{\sim}{Y} + \underset{\sim}{Y}^{\dagger} \geq 0 \tag{8.58}$$

where we mean in setting a matrix greater than or equal to zero, that it be positive definite [43, Sec. 1.17]. Similarly, from Eq. 8.50, the scattering matrix describing a passive system fulfills the condition

$$\underset{\sim}{1} - \underset{\sim}{S}^{\dagger}\underset{\sim}{S} \geq 0 \tag{8.59}$$

And finally, a network composed of only resistances and inductances has a positive reactive power input, so from Eq. 8.51

$$j(\underset{\sim}{Z}^{\dagger} - \underset{\sim}{Z}) \geq 0 \tag{8.60}$$

and

$$j(\underset{\sim}{Y} - \underset{\sim}{Y}^{\dagger}) \geq 0 \tag{8.61}$$

Similarly, a capacitive network has a negative Q, so

$$j(\underset{\sim}{Z} - \underset{\sim}{Z}^{\dagger}) \geq 0 \tag{8.62}$$

and

$$j(\underset{\sim}{Y}^{\dagger} - \underset{\sim}{Y}) \geq 0 \tag{8.63}$$

These constraints for special types of linear time-invariant circuits are given in Table 8.1. The constraints are well known; they were derived here primarily to indicate the method used in the next section for networks obeying one of the four types of frequency-power formulas.

Linear Time-Varying Matrix Constraints. Suppose a linear parametrically-pumped system, of the type described in Sec. 8.2, has a

Lossless	$P = 0$	$\underset{\sim}{Z} = -\underset{\sim}{Z}^\dagger$ $\underset{\sim}{Y} = -\underset{\sim}{Y}^\dagger$ $\underset{\sim}{S}^{-1} = \underset{\sim}{S}^\dagger$
Pure Lossy	$Q = 0$	$\underset{\sim}{Z} = \underset{\sim}{Z}^\dagger$ $\underset{\sim}{Y} = \underset{\sim}{Y}^\dagger$
Passive	$P \geq 0$	$\underset{\sim}{Z} + \underset{\sim}{Z}^\dagger \geq 0$ $\underset{\sim}{Y} + \underset{\sim}{Y}^\dagger \geq 0$ $\underset{\sim}{1} - \underset{\sim}{S}^\dagger \underset{\sim}{S} \geq 0$ $\underset{\sim}{1} - \underset{\sim}{S}\underset{\sim}{S}^\dagger \geq 0$
Inductive	$Q \geq 0$	$j(\underset{\sim}{Z}^\dagger - \underset{\sim}{Z}) \geq 0$ $j(\underset{\sim}{Y} - \underset{\sim}{Y}^\dagger) \geq 0$
Capacitive	$Q \leq 0$	$j(\underset{\sim}{Z} - \underset{\sim}{Z}^\dagger) \geq 0$ $j(\underset{\sim}{Y}^\dagger - \underset{\sim}{Y}) \geq 0$

Table 8.1. Matrix constraints for linear time-invariant networks of various types

finite number of ports, as shown in Fig. 8.2. In general, each port exchanges power at a number of frequencies. Let us represent this system by Fig. 8.3, where we show extra terminal pairs, so that each terminal pair now exchanges energy at only one frequency.

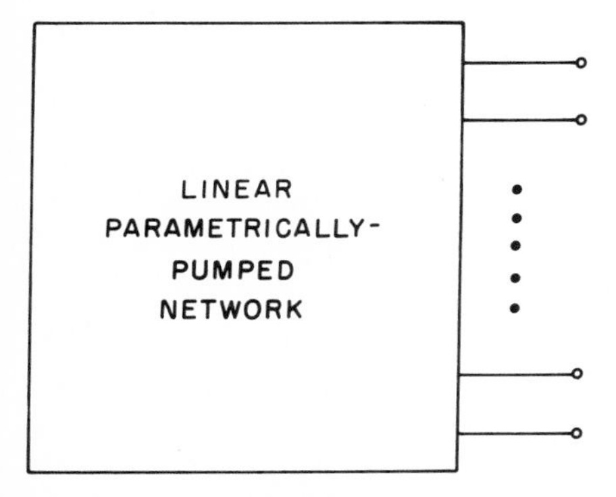

Fig. 8.2. A linear parametrically-pumped system, with a small number of physical ports. Each port exchanges power at a number of frequencies

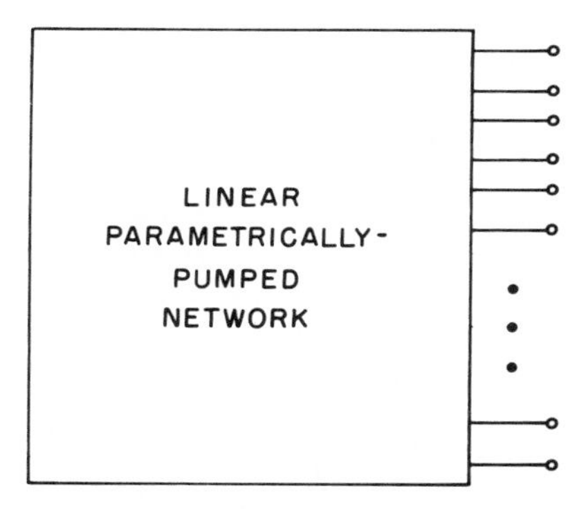

Fig. 8.3. The same system shown in Fig. 8.2, but with additional fictitious ports added, so that each new port exchanges power at only one frequency

This is a nonphysical representation for the network, in the sense that one physical element may appear more than once if currents of various frequencies flow through it. In addition, the loads placed on the ports are not all physically separate, so any given physical element may appear in several of them.

For example, consider the network of Fig. 8.4, which is a reasonably good approximation of a parametrically-pumped varactor [107], including the series resistance. This physical device has only one terminal pair, but when it is used as a frequency converter it is convenient to represent it as a poly-port network. The network of Fig. 8.5 would be suitable if power is exchanged at only two frequencies. Note that the series resistance appears at both terminal pairs, even though it is only one physical element.

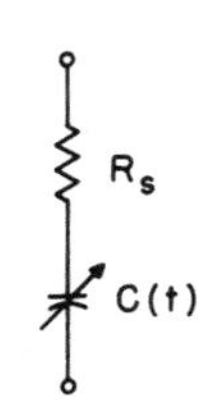

Fig. 8.4. Model for a parametrically pumped varactor, including series resistance

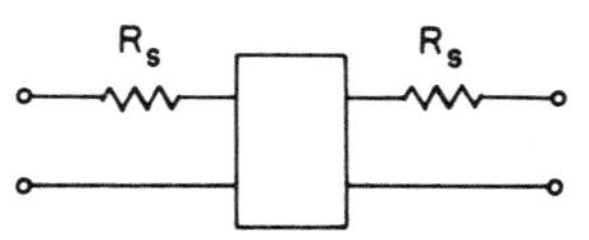

Fig. 8.5. A schematic representation for the parametrically-pumped varactor. Each port shown is at a different frequency; the voltage and current at each port are merely the components of actual varactor voltage and current at the corresponding frequency. Note that because the series resistance is linear, it may be put separately in each lead, so that the box shown consists only of the varying capacitance, and therefore satisfies frequency-power formulas of Types I and IV

Assume the pumping is at frequency ω_p, and the small signals at frequencies $(k\omega_p + \omega_s)$. Because the system of Fig. 8.3 is linear, the equations relating the terminal variables are linear and can be put in matrix form. Each terminal pair has voltage and current with frequency ω_i (of the form $k\omega_p + \omega_s$) and amplitudes E_i and I_i. Then we may define $\underset{\sim}{E}$ and $\underset{\sim}{I}$ as column matrices of the E_i and I_i, and write the "conversion matrices" $\underset{\sim}{Z}$ and $\underset{\sim}{Y}$ such that [106, Ch. 5]

$$\underset{\sim}{E} = \underset{\sim}{Z}\underset{\sim}{I} \tag{8.64}$$

and

$$\underset{\sim}{I} = \underset{\sim}{Y}\underset{\sim}{E} \tag{8.65}$$

Alternatively, we can define the scattering variables at each port by Eqs. 8.45 and 8.46, and write the equations of motion in the form

$$\underset{\sim}{b} = \underset{\sim}{S}\underset{\sim}{a} \tag{8.66}$$

using the scattering matrix $\underset{\sim}{S}$. Although $\underset{\sim}{Z}$, $\underset{\sim}{Y}$, and $\underset{\sim}{S}$ describe a time-varying system, they do not vary in time themselves.

If the network of Fig. 8.3 obeys frequency-power formulas of any type, these place constraints on the matrices $\underset{\sim}{Z}$, $\underset{\sim}{Y}$, and $\underset{\sim}{S}$, similar to the constraints of Table 8.1. We now calculate these matrix constraints, and tabulate them.

Suppose the network of Fig. 8.3 consists of reactive-type elements, gyrators, transformers, and other devices that obey the Manley-Rowe formulas. Then we write, from Eq. 8.37,

$$\Sigma_i \frac{\frac{1}{2}\mathrm{Re}(E_i I_i^*)}{\omega_i} = 0 \tag{8.67}$$

where the sum over frequencies is replaced by a sum over ports, because each port passes power at only one frequency. In terms of the diagonal matrix $\underset{\sim}{K}$ with entries $1/\omega_i$ along the main diagonal, Eq. 8.67 may be written

$$\tfrac{1}{4}(\underset{\sim}{I}\dagger\underset{\sim}{K}\underset{\sim}{E} + \underset{\sim}{E}\dagger\underset{\sim}{K}\underset{\sim}{I}) = 0 \tag{8.68}$$

or

$$\begin{aligned} 0 &= \underset{\sim}{I}\dagger(\underset{\sim}{K}\underset{\sim}{Z} + \underset{\sim}{Z}\dagger\underset{\sim}{K})\underset{\sim}{I} \\ &= \underset{\sim}{E}\dagger(\underset{\sim}{Y}\dagger\underset{\sim}{K} + \underset{\sim}{K}\underset{\sim}{Y})\underset{\sim}{E} \end{aligned} \tag{8.69}$$

and because $\underset{\sim}{I}$ and $\underset{\sim}{E}$ can be arbitrary in Eq. 8.69,

$$\underset{\sim}{K}\underset{\sim}{Z} + \underset{\sim}{Z}\dagger\underset{\sim}{K} = 0 \tag{8.70}$$

and

$$\underset{\sim}{K}\underset{\sim}{Y} + \underset{\sim}{Y}\dagger\underset{\sim}{K} = 0 \tag{8.71}$$

In addition, Eq. 8.67 can be written

$$\Sigma_i \frac{|a_i|^2 - |b_i|^2}{\omega_i} = 0 \tag{8.72}$$

using the scattering variables, so

$$\begin{aligned} 0 &= \underset{\sim}{a}\dagger\underset{\sim}{K}\underset{\sim}{a} - \underset{\sim}{b}\dagger\underset{\sim}{K}\underset{\sim}{b} \\ &= \underset{\sim}{a}\dagger(\underset{\sim}{K} - \underset{\sim}{S}\dagger\underset{\sim}{K}\underset{\sim}{S})\underset{\sim}{a} \end{aligned} \tag{8.73}$$

whereupon we conclude that

$$\underset{\sim}{K} = \underset{\sim}{S}\dagger\underset{\sim}{K}\underset{\sim}{S} \tag{8.74}$$

Equations 8.70, 8.71, and 8.74 are the constraints the Manley-Rowe formulas place on the system matrices $\underset{\sim}{Z}$, $\underset{\sim}{Y}$, and $\underset{\sim}{S}$. Many other equivalent forms can be obtained from these constraints; for example, premultiply Eq. 8.74 by $\underset{\sim}{K}^{-1}\underset{\sim}{S}\dagger^{-1}$ and postmultiply it by $\underset{\sim}{K}^{-1}\underset{\sim}{S}\dagger$, to obtain

$$\underset{\sim}{S}\underset{\sim}{K}^{-1}\underset{\sim}{S}\dagger = \underset{\sim}{K}^{-1} \tag{8.75}$$

a form that has been used before [40]. In Table 8.2 we list several variations of the constraints of Eqs. 8.70, 8.71, and 8.74.

Type I (Manley-Rowe Formulas)	$\underset{\sim}{K}\underset{\sim}{Z} + \underset{\sim}{Z}\dagger\underset{\sim}{K} = 0$	$\underset{\sim}{K}\underset{\sim}{Y} + \underset{\sim}{Y}\dagger\underset{\sim}{K} = 0$
	$\underset{\sim}{Z} = -\underset{\sim}{K}^{-1}\underset{\sim}{Z}\dagger\underset{\sim}{K}$	$\underset{\sim}{Y} = -\underset{\sim}{K}^{-1}\underset{\sim}{Y}\dagger\underset{\sim}{K}$
	$\underset{\sim}{Z}\dagger = -\underset{\sim}{K}\underset{\sim}{Z}\underset{\sim}{K}^{-1}$	$\underset{\sim}{Y}\dagger = -\underset{\sim}{K}\underset{\sim}{Y}\underset{\sim}{K}^{-1}$
	$(\underset{\sim}{Z})^n = (-1)^n\underset{\sim}{K}^{-1}(\underset{\sim}{Z}\dagger)^n\underset{\sim}{K}$	$(\underset{\sim}{Y})^n = (-1)^n\underset{\sim}{K}^{-1}(\underset{\sim}{Y}\dagger)^n\underset{\sim}{K}$
	$(\underset{\sim}{Z}\dagger)^n = (-1)^n\underset{\sim}{K}(\underset{\sim}{Z})^n\underset{\sim}{K}^{-1}$	$(\underset{\sim}{Y}\dagger)^n = (-1)^n\underset{\sim}{K}(\underset{\sim}{Y})^n\underset{\sim}{K}^{-1}$
	$\underset{\sim}{K} = -\underset{\sim}{Z}\dagger\underset{\sim}{K}\underset{\sim}{Y}$	$\underset{\sim}{K}^{-1} = -\underset{\sim}{Z}\underset{\sim}{K}^{-1}\underset{\sim}{Y}\dagger$
	$= -\underset{\sim}{Y}\dagger\underset{\sim}{K}\underset{\sim}{Z}$	$= -\underset{\sim}{Y}\underset{\sim}{K}^{-1}\underset{\sim}{Z}\dagger$
	$= (-1)^n(\underset{\sim}{Z}\dagger)^n\underset{\sim}{K}(\underset{\sim}{Y})^n$	$= (-1)^n(\underset{\sim}{Z})^n\underset{\sim}{K}^{-1}(\underset{\sim}{Y}\dagger)^n$
	$= (-1)^n(\underset{\sim}{Y}\dagger)^n\underset{\sim}{K}(\underset{\sim}{Z})^n$	$= (-1)^n(\underset{\sim}{Y})^n\underset{\sim}{K}^{-1}(\underset{\sim}{Z}\dagger)^n$
	$\underset{\sim}{K} = \underset{\sim}{S}\dagger\underset{\sim}{K}\underset{\sim}{S}$	$\underset{\sim}{K}^{-1} = \underset{\sim}{S}\underset{\sim}{K}^{-1}\underset{\sim}{S}\dagger$
	$= (\underset{\sim}{S}\dagger)^n\underset{\sim}{K}(\underset{\sim}{S})^n$	$= (\underset{\sim}{S})^n\underset{\sim}{K}^{-1}(\underset{\sim}{S}\dagger)^n$
	$\underset{\sim}{S} = \underset{\sim}{K}^{-1}(\underset{\sim}{S}\dagger)^{-1}\underset{\sim}{K}$	$\underset{\sim}{S}\dagger = \underset{\sim}{K}\underset{\sim}{S}^{-1}\underset{\sim}{K}^{-1}$
	$(\underset{\sim}{S})^n = \underset{\sim}{K}^{-1}(\underset{\sim}{S}\dagger)^{-n}\underset{\sim}{K}$	$(\underset{\sim}{S}\dagger)^n = \underset{\sim}{K}(\underset{\sim}{S})^{-n}\underset{\sim}{K}^{-1}$
Type II	$\underset{\sim}{Y} = \underset{\sim}{Y}\dagger$	$\underset{\sim}{Z} = \underset{\sim}{Z}\dagger$
Type III	$\underset{\sim}{Z} + \underset{\sim}{Z}\dagger \geq 0$	$\underset{\sim}{1} - \underset{\sim}{S}\underset{\sim}{S}\dagger \geq 0$
	$\underset{\sim}{Y} + \underset{\sim}{Y}\dagger \geq 0$	$\underset{\sim}{1} - \underset{\sim}{S}\dagger\underset{\sim}{S} \geq 0$
Type IV Inductive	$j(\underset{\sim}{Z}\dagger\underset{\sim}{K} - \underset{\sim}{K}\underset{\sim}{Z}) \geq 0$	$j(\underset{\sim}{K}\underset{\sim}{Y} - \underset{\sim}{Y}\dagger\underset{\sim}{K}) \geq 0$
Type IV Capacitive	$j(\underset{\sim}{K}\underset{\sim}{Z} - \underset{\sim}{Z}\dagger\underset{\sim}{K}) \geq 0$	$j(\underset{\sim}{Y}\dagger\underset{\sim}{K} - \underset{\sim}{K}\underset{\sim}{Y}) \geq 0$
Types I and IV Inductive	$j\underset{\sim}{Z}\dagger\underset{\sim}{K} = -j\underset{\sim}{K}\underset{\sim}{Z} \geq 0$	$j\underset{\sim}{K}\underset{\sim}{Y} = -j\underset{\sim}{Y}\dagger\underset{\sim}{K} \geq 0$
Types I and IV Capacitive	$j\underset{\sim}{K}\underset{\sim}{Z} = -j\underset{\sim}{Z}\dagger\underset{\sim}{K} \geq 0$	$j\underset{\sim}{Y}\dagger\underset{\sim}{K} = -j\underset{\sim}{K}\underset{\sim}{Y} \geq 0$
Types II and III	$\underset{\sim}{Z} = \underset{\sim}{Z}\dagger \geq 0$	$\underset{\sim}{Y} = \underset{\sim}{Y}\dagger \geq 0$

Table 8.2. Various forms of the matrix constraints for linear time-varying circuits that obey one or more frequency-power formulas; n is a positive or negative integer

Now suppose the network of Fig. 8.3 obeys frequency-power formulas of the second type. Then by Eq. 8.38 the sum of the reactive powers in at each port vanishes; thus

$$0 = \Sigma_i Q_i = \frac{1}{4j} \underset{\sim}{E}^\dagger (\underset{\sim}{Y}^\dagger - \underset{\sim}{Y}) \underset{\sim}{E}$$

$$= \frac{1}{4j} \underset{\sim}{I}^\dagger (\underset{\sim}{Z} - \underset{\sim}{Z}^\dagger) \underset{\sim}{I} \tag{8.76}$$

and hence

$$\underset{\sim}{Z} = \underset{\sim}{Z}^\dagger \tag{8.77}$$

and

$$\underset{\sim}{Y} = \underset{\sim}{Y}^\dagger \tag{8.78}$$

so both $\underset{\sim}{Z}$ and $\underset{\sim}{Y}$ are Hermitian. Note that some of the frequencies ω_i may be negative, and for these E_i and I_i are the complex conjugates of the amplitudes of the corresponding positive frequency, and the matrices $\underset{\sim}{Z}$ and $\underset{\sim}{Y}$ must be written accordingly, for Eqs. 8.77 and 8.78 to hold.

Now suppose the network of Fig. 8.3 obeys the third type of frequency-power formulas. Then Eq. 8.39 predicts that the total real power input is positive. Just as for passive networks, this condition predicts that

$$\underset{\sim}{Z} + \underset{\sim}{Z}^\dagger \geq 0 \tag{8.79}$$

$$\underset{\sim}{Y} + \underset{\sim}{Y}^\dagger \geq 0 \tag{8.80}$$

and

$$\underset{\sim}{1} - \underset{\sim}{S}^\dagger \underset{\sim}{S} \geq 0 \tag{8.81}$$

Now suppose the network of Fig. 8.3 obeys the fourth type of frequency-power formulas, and is inductive. Then, from Eq. 8.40,

$$\Sigma_i \frac{Q_i}{\omega_i} \geq 0 \tag{8.82}$$

and the matrix constraints are

$$j(\underset{\sim}{Z}^\dagger \underset{\sim}{K} - \underset{\sim}{K}\underset{\sim}{Z}) \geq 0 \tag{8.83}$$

and

$$j(\underset{\sim}{K}\underset{\sim}{Y} - \underset{\sim}{Y}^\dagger \underset{\sim}{K}) \geq 0 \tag{8.84}$$

On the other hand, if the system obeys these formulas but is capacitive, the matrix constraints become, from Eq. 8.41,

$$j(\underset{\sim}{K}\underset{\sim}{Z} - \underset{\sim}{Z}^{\dagger}\underset{\sim}{K}) \geq 0 \tag{8.85}$$

and

$$j(\underset{\sim}{Y}^{\dagger}\underset{\sim}{K} - \underset{\sim}{K}\underset{\sim}{Y}) \geq 0 \tag{8.86}$$

We have found matrix constraints imposed by each of the four types of frequency-power formulas. These are summarized in Table 8.2.

Many devices obey more than one type, however, and it is interesting to note the constraints that apply to them. If a device satisfies formulas of Types I and IV simultaneously, then, if it is inductive, the matrices $-j\underset{\sim}{K}\underset{\sim}{Z}$ and $j\underset{\sim}{K}\underset{\sim}{Y}$ are Hermitian and positive definite; if it is capacitive, the matrices $j\underset{\sim}{K}\underset{\sim}{Z}$ and $-j\underset{\sim}{K}\underset{\sim}{Y}$ are Hermitian and positive definite. Similarly, if a system obeys frequency-power formulas of both Types II and III, then $\underset{\sim}{Z}$ and $\underset{\sim}{Y}$ are Hermitian and positive definite.

8.4 Impedance Translation

It is our intent here to mention briefly one typical application involving frequency-power formulas of more than one type. We do not propose to discuss it thoroughly, but only to show that the frequency-power formulas can be used to gain insight and even to obtain definite fundamental limits on performance.

Suppose a one-port linear parametrically-pumped network is connected to a passive load Z_n at one frequency, $n\omega_p + \omega_s$, where ω_p is the pump frequency and ω_s is another frequency, and suppose that the external network is adjusted so that power flows only at frequencies $n\omega_p + \omega_s$ and ω_s. Then, by the method of Sec. 8.3, we represent this system by Fig. 8.6, with a terminal pair for each frequency. Our desire is to determine the input impedance at the ω_s terminal pair.

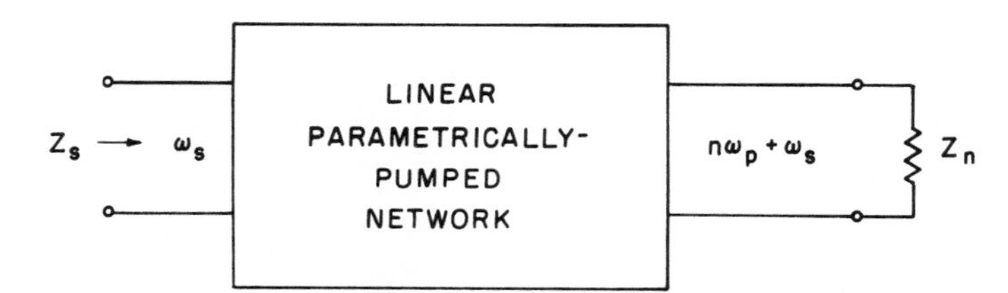

Fig. 8.6. A linear parametrically-pumped one-port system exchanging power at only two frequencies. This device is suitable for impedance translation

This device might be used to translate an impedance from one frequency to another, for example to obtain an impedance function at frequencies close to ω_s, that cannot be realized directly with the present state of the art, using convenient elements. Do the frequency-power formulas indicate any fundamental limits on the performance of such a device?

At least two applications of this idea have been made. A band-pass filter using a similar technique was described by Franks and Witt [27]; whereas the simple circuit of Fig. 8.6 does not describe their apparatus, it can do so if more terminals are added. And secondly, a striking example is the parametric amplifier, in which a passive idler impedance Z_n is translated to an impedance at ω_s with a negative real part!

To see in a simple example how the frequency-power formulas might predict fundamental limits of operation for such devices, suppose the parametrically-pumped network consists of nonlinear resistors, so that it obeys frequency-power formulas of Types II and III. Then, if P_s and Q_s are the real and reactive power flowing into the network at frequency ω_s, and if P_n and Q_n are the real and reactive power flowing <u>into the impedance</u> Z_n, the frequency-power formulas predict

$$P_s \geq P_n \tag{8.87}$$

and

$$Q_s = Q_n \tag{8.88}$$

These conditions constrain the possible impedances Z_s we may observe at the left-hand terminals, for

$$\begin{aligned} P_s + jQ_s &= \tfrac{1}{2} E_s I_s^* \\ &= \tfrac{1}{2} |I_s|^2 Z_s \end{aligned} \tag{8.89}$$

and

$$\begin{aligned} P_n + jQ_n &= \tfrac{1}{2} E_n I_n^* \\ &= \tfrac{1}{2} |I_n|^2 Z_n \end{aligned} \tag{8.90}$$

Thus the real and imaginary parts of Z_s are restricted, for

$$\begin{aligned} \operatorname{Re} Z_s &= \frac{2P_s}{|I_s|^2} \\ &\geq \frac{2P_n}{|I_s|^2} = \left| \frac{I_n}{I_s} \right|^2 \operatorname{Re} Z_n \end{aligned} \tag{8.91}$$

and

$$\operatorname{Im} Z_s = \frac{2Q_s}{|I_s|^2}$$

$$= \frac{2Q_n}{|I_s|^2} = \left|\frac{I_n}{I_s}\right|^2 \operatorname{Im} Z_n \tag{8.92}$$

The phase angle of Z_s must therefore lie between zero and the phase angle of Z_n. If we mark the point Z_n in the complex plane, as in Fig. 8.7, the region of possible values of Z_s is that region to the right of the ray passing through Z_n, or the shaded area of Fig. 8.7.

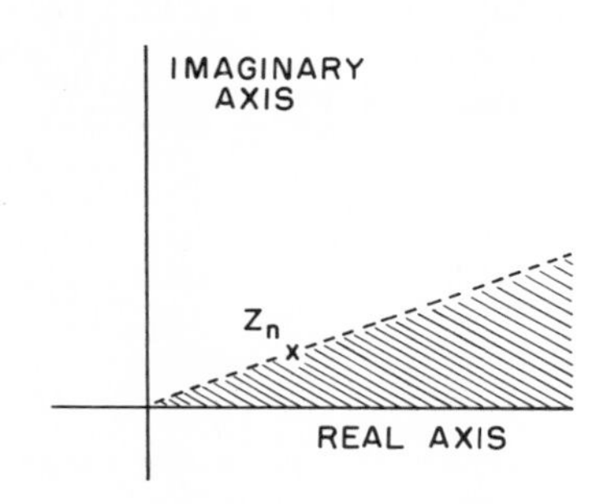

Fig. 8.7. The complex plane, showing the impedance Z_n. The frequency-power formulas show that Z_s is restricted to lie in the shaded region

An interesting result of this is that if the impedance Z_n is that of a tuned circuit with a certain quality factor, the quality factor of the translated impedance function near frequency ω_s is less than or at best equal to the original quality factor multiplied by the ratio $\omega_s/(n\omega_p+\omega_s)$.

These results are probably not important in themselves, but they do serve as an indication of the ability of the frequency-power formulas to predict fundamental limits.

APPENDIX A

We show here, with a counterexample, that not all lossless devices obey the Manley-Rowe formulas; that is, the frequency-power formulas of the first type. The device is a simple switch, thrown periodically with a fundamental frequency ω_p, and connected to a voltage source $E \cos \omega_s t$ (with $\omega_s < \omega_p$) and a load resistor R. The circuit is shown in Fig. A.1. According to Sec. 5.2, the switch does obey the second and third types of frequency-power formula, but we show by direct calculation that it does not satisfy the first type.

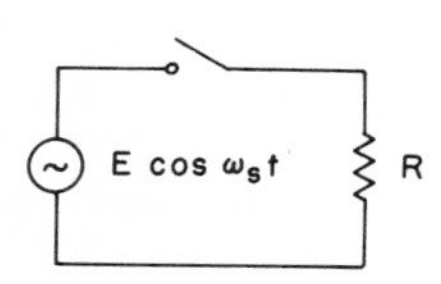

Fig. A.1. Circuit used to show that a switch does not obey the Manley-Rowe formulas, even though it is lossless

The current through the switch is $(E/R)\cos \omega_s t$ when the switch is closed and 0 when the switch is open. Thus if $\theta(t)$ is a function that is 1 when the switch is closed and 0 when the switch is open, the current $i(t)$ is

$$i(t) = \theta(t) \frac{E}{R} \cos \omega_s t \tag{A.1}$$

and the switch voltage $e(t)$ is

$$e(t) = [1 - \theta(t)]\, E \cos \omega_s t \tag{A.2}$$

If the switch is open half the time and closed half the time, and is closed during a region symmetrical about the origin, a simple Fourier analysis shows

$$\theta(t) = \tfrac{1}{2} + \sum_{n=1}^{\infty} \theta_n \cos n\omega_p t \tag{A.3}$$

where

$$\theta_n = \begin{cases} \dfrac{2}{n\pi} (-1)^{(n-1)/2} & n \text{ odd} \\ 0 & n \text{ even} \end{cases} \tag{A.4}$$

Thus

$$i(t) = \frac{E}{2R} \cos \omega_s t + \sum_{n=1}^{\infty} \frac{E\theta_n}{2R} \cos(n\omega_p + \omega_s)t + \sum_{n=1}^{\infty} \frac{E\theta_n}{2R} \cos(n\omega_p - \omega_s)t \tag{A.5}$$

and

$$e(t) = \frac{E}{2} \cos \omega_s t - \sum_{n=1}^{\infty} \frac{E}{2} \theta_n \cos(n\omega_p + \omega_s)t - \sum_{n=1}^{\infty} \frac{E}{2} \theta_n \cos(n\omega_p - \omega_s)t \tag{A.6}$$

Thus power flows into the switch at frequency ω_s of magnitude

$$P_s = \frac{E^2}{8R} \tag{A.7}$$

and out at all frequencies of the form $(n\omega_p \pm \omega_s)$ of magnitude

$$-P_{np\pm s} = \frac{E^2}{8R}\theta_n{}^2$$

$$= \begin{cases} \dfrac{E^2}{2n^2\pi^2 R} & \text{(n odd)} \\ 0 & \text{(n even)} \end{cases} \tag{A.8}$$

The reactive power input at each frequency is 0.

Choosing ω_s and ω_p as independent frequencies, the first type of frequency-power formula would predict

$$\frac{P_s}{\omega_s} + \sum_{n=1}^{\infty}\left[\frac{P_{np+s}}{n\omega_p + \omega_s} - \frac{P_{np-s}}{n\omega_p - \omega_s}\right] = 0 \tag{A.9}$$

and possibly

$$\sum_{n=1}^{\infty}\left[\frac{nP_{np+s}}{n\omega_p + \omega_s} + \frac{nP_{np-s}}{n\omega_p - \omega_s}\right] = 0 \tag{A.10}$$

neither of which is true.

Note, however, that the device does obey the second and third types of frequency-power formulas, as we would expect because its operation is defined in terms of its voltage and current. It is the limit of a time-varying resistor. Because the time-variation proceeds at frequency ω_p (and its harmonics), there is only one frequency-power formula of Type II:

$$\sum_{n=1}^{\infty} Q_{np+s} - \sum_{n=1}^{\infty} Q_{np-s} + Q_s = 0 \tag{A.11}$$

and only one of Type III:

$$\sum_{n=1}^{\infty} P_{np+s} + \sum_{n=1}^{\infty} P_{np-s} + P_s \geq 0 \tag{A.12}$$

Each of these is evidently consistent with the actual results (see Eqs. A.7 and A.8, and the remark following Eq. A.8).

Other lossless devices that are known not to obey the Manley-Rowe formulas include the ideal diode, variable transformer, and their mechanical counterparts.

APPENDIX B

Here we tabulate for convenience the four types of frequency-power formulas. In Table B.1 the general formulas are given, without any particular frequency constraints. In Table B.2 these formulas are specialized to the case of most practical interest, for frequencies of the form $m\omega_1 + n\omega_0$. If only frequencies of the form $n\omega$ are present, the formulas assume the expressions of Table B.3. Linear time-varying systems with frequencies of the form $n\omega_p + \omega_s$, where n is positive or negative, might obey the formulas in Table B.4, where we assume that ω_p is the pump frequency, and is the only frequency that enters into the explicit time dependence of the system; there is only one formula of each type.

If only a few frequencies are present, the formulas are very simple. Table B.5 lists some of the formulas for a three-frequency system, and Table B.6 for a four-frequency system.

In using these tables, care must be taken in interpreting the Q_α as reactive powers when the associated ω_α are negative; see the remarks following Eq. 3.26.

Table B.7 contains a list of some of the devices that are known to obey the formulas of each type; the list is, of course, not complete. A few of the more important devices that are known <u>not</u> to obey formulas of each type are listed in Table B.8.

Table B.1. Frequency-power formulas of the four types, under general frequency constraints. The formulas hold for any values of the parameters θ_a and χ_a.

Type I Manley-Rowe Formulas (with d-c power)	$\Sigma_i \frac{P_{i0}}{v_{i0}} \frac{\partial v_{i0}}{\partial \omega_a} + \Sigma_\alpha \frac{P_\alpha}{\omega_\alpha} \frac{\partial \omega_\alpha}{\partial \omega_a} = 0$
Type I Manley-Rowe Formulas (without d-c power)	$\Sigma_\alpha \frac{P_\alpha}{\omega_\alpha} \frac{\partial \omega_\alpha}{\partial \omega_a} = 0$
Type II	$\Sigma_\alpha \frac{\partial \omega_\alpha}{\partial \omega_a} Q_\alpha = 0$
Type III (Page form)	$\Sigma_\alpha P_\alpha \left(1 - \cos \frac{\partial \omega_\alpha}{\partial \omega_a} \theta_a\right) \geq 0$
	$\Sigma_\alpha P_\alpha \left[1 - \cos\left(\Sigma_a \frac{\partial \omega_\alpha}{\partial \omega_a} \theta_a\right)\right] \geq 0$
(Pantell form)	$\Sigma_\alpha P_\alpha \left(\frac{\partial \omega_\alpha}{\partial \omega_a}\right)^2 \geq 0$
	$\Sigma_\alpha P_\alpha \left(\Sigma_a \frac{\partial \omega_\alpha}{\partial \omega_a} \chi_a\right)^2 \geq 0$
	$\Sigma_\alpha P_\alpha g_\alpha(\theta_a) \geq 0$

Table B. 1. (cont.)

Type IV Inductive	$\Sigma_\alpha \frac{Q_\alpha}{\omega_\alpha}\left(1 - \cos \frac{\partial\omega_\alpha}{\partial\omega_a}\, \theta_a\right) \geq 0$ $\Sigma_\alpha \frac{Q_\alpha}{\omega_\alpha}\left[1 - \cos\left(\Sigma_a \frac{\partial\omega_\alpha}{\partial\omega_a}\, \theta_a\right)\right] \geq 0$ $\Sigma_\alpha \frac{Q_\alpha}{\omega_\alpha}\left(\frac{\partial\omega_\alpha}{\partial\omega_a}\right)^2 \geq 0$ $\Sigma_\alpha \frac{Q_\alpha}{\omega_\alpha}\left(\Sigma_a \frac{\partial\omega_\alpha}{\partial\omega_a}\, \chi_a\right)^2 \geq 0$ $\Sigma_\alpha \frac{Q_\alpha}{\omega_\alpha}\, g_\alpha(\theta_a) \geq 0$
Type IV Capacitive	$\Sigma_\alpha \frac{Q_\alpha}{\omega_\alpha}\left(1 - \cos \frac{\partial\omega_\alpha}{\partial\omega_a}\, \theta_a\right) \leq 0$ $\Sigma_\alpha \frac{Q_\alpha}{\omega_\alpha}\left[1 - \cos\left(\Sigma_a \frac{\partial\omega_\alpha}{\partial\omega_a}\, \theta_a\right)\right] \leq 0$ $\Sigma_\alpha \frac{Q_\alpha}{\omega_\alpha}\left(\frac{\partial\omega_\alpha}{\partial\omega_a}\right)^2 \leq 0$ $\Sigma_\alpha \frac{Q_\alpha}{\omega_\alpha}\left(\Sigma_a \frac{\partial\omega_\alpha}{\partial\omega_a}\, \chi_a\right)^2 \leq 0$ $\Sigma_\alpha \frac{Q_\alpha}{\omega_\alpha}\, g_\alpha(\theta_a) \leq 0$
where $g_\alpha(\theta_a) = \left(\Sigma_a \frac{\partial\omega_\alpha}{\partial\omega_a}\, \theta_a\right) \int_0^\infty \xi(x) \sin\left(\Sigma_a \frac{\partial\omega_\alpha}{\partial\omega_a}\, \theta_a x\right) dx$ with $\xi(x)$ any nonincreasing positive function defined from **0** to $+\infty$.	

Table B.2. Frequency-power formulas of the four types, with frequencies of the form $m\omega_1 + n\omega_0$. The formulas of types III and IV hold for all values of θ_1, θ_0, χ_1 and χ_0.

Type I; Manley-Rowe Formulas (without d-c power)	$\sum_m \sum_n \frac{m P_{mn}}{m\omega_1 + n\omega_0} = 0$ $\sum_m \sum_n \frac{n P_{mn}}{m\omega_1 + n\omega_0} = 0$
Type II	$\sum_m \sum_n m Q_{mn} = 0$ $\sum_m \sum_n n Q_{mn} = 0$
Type III	$\sum_m \sum_n P_{mn} [1 - \cos(m\theta_1 + n\theta_0] \geq 0$ $\sum_m \sum_n P_{mn} (m\chi_1 + n\chi_0)^2 \geq 0$ $\sum_m \sum_n P_{mn} g_{mn} (\theta_1, \theta_0) \geq 0$
Type IV Inductive	$\sum_m \sum_n \frac{Q_{mn} [1 - \cos(m\theta_1 + n\theta_0)]}{m\omega_1 + n\omega_0} \geq 0$ $\sum_m \sum_n \frac{Q_{mn} (m\chi_1 + n\chi_0)^2}{m\omega_1 + n\omega_0} \geq 0$ $\sum_m \sum_n \frac{Q_{mn} g_{mn} (\theta_1, \theta_0)}{m\omega_1 + n\omega_0} \geq 0$
Type IV Capacitive	$\sum_m \sum_n \frac{Q_{mn} [1 - \cos(m\theta_1 + n\theta_0)]}{m\omega_1 + n\omega_0} \leq 0$ $\sum_m \sum_n \frac{Q_{mn} (m\chi_1 + n\chi_0)^2}{m\omega_1 + n\omega_0} \leq 0$ $\sum_m \sum_n \frac{Q_{mn} g_{mn} (\theta_1, \theta_0)}{m\omega_1 + n\omega_0} \leq 0$

where

$$g_{mn}(\theta_1, \theta_0) = (m\theta_1 + n\theta_0) \int_0^\infty \xi(x) \sin(m\theta_1 x + n\theta_0 x)\, dx$$

where $\xi(x)$ is any nonincreasing positive function defined from 0 to $+\infty$.

Table B.3. Frequency-power formulas of the four types, with frequencies of the form $n\omega$. The formulas of Types III and IV hold for all values of θ.

Type I Manley-Rowe formulas	$\Sigma_n P_n = 0$
Type II	$\Sigma_n n Q_n = 0$
Type III	$\Sigma_n P_n (1 - \cos n\theta) \geq 0$ $\Sigma_n n^2 P_n \geq 0$ $\Sigma_n P_n g_n(\theta) \geq 0$
Type IV Inductive	$\Sigma_n \frac{Q_n}{n} (1 - \cos n\,\theta) \geq 0$ $\Sigma_n n Q_n \geq 0$ $\Sigma_n \frac{Q_n}{n} g_n(\theta) \geq 0$
Type IV Capacitive	$\Sigma_n \frac{Q_n}{n} (1 - \cos n\,\theta) \leq 0$ $\Sigma_n n Q_n \leq 0$ $\Sigma_n \frac{Q_n}{n} g_n(\theta) \leq 0$

where

$$g_n(\theta) = n\,\theta \int_0^\infty \xi(x) \sin(n\theta\, x)\,dx$$

with $\xi(x)$ any nonincreasing positive function defined from 0 to ∞.

Table B.4. Frequency-power formulas of the four types, with frequencies of the form $n\omega_p + \omega_s$; it is assumed that ω_p specifies the explicit time dependence of the system

Type I Manley-Rowe Formulas	$\Sigma_n \dfrac{P_n}{n\omega_p + \omega_s} = 0$
Type II	$\Sigma_n Q_n = 0$
Type III	$\Sigma_n P_n \geq 0$
Type IV Inductive	$\Sigma_n \dfrac{Q_n}{n\omega_p + \omega_s} \geq 0$
Type IV Capacitive	$\Sigma_n \dfrac{Q_n}{n\omega_p + \omega_s} \leq 0$

Table B.5. Frequency-power formulas of the four types, with only three frequencies: ω_s, ω_p, and $\omega_- = \omega_p - \omega_s$. Only a few of the possible formulas of Types III and and IV are given; they hold for all values of the parameters θ_s, θ_p, χ_s, and χ_p

Type I Manley-Rowe Formulas	$\frac{P_s}{\omega_s} = \frac{P_-}{\omega_-} = -\frac{P_p}{\omega_p}$
Type II	$Q_s = Q_- = -Q_p$
Type III	$P_s[1 - \cos\theta_s] + P_p[1 - \cos\theta_p] + P_-[1 - \cos(\theta_p - \theta_s)] \geq 0$ $P_s\chi_s^2 + P_p\chi_p^2 + P_-(\chi_p - \chi_s)^2 \geq 0$ $P_s + P_- \geq 0$ $P_s + P_p \geq 0$ $P_p + P_- \geq 0$
Type IV Inductive	$\frac{Q_s[1 - \cos\theta_s]}{\omega_s} + \frac{Q_p[1 - \cos\theta_p]}{\omega_p} + \frac{Q_-[1 - \cos(\theta_p - \theta_s)]}{\omega_-} \geq 0$ $\frac{Q_s\chi_s^2}{\omega_s} + \frac{Q_p\chi_p^2}{\omega_p} + \frac{Q_-(\chi_p - \chi_s)^2}{\omega_-} \geq 0$ $\frac{Q_s}{\omega_s} + \frac{Q_-}{\omega_-} \geq 0$ $\frac{Q_s}{\omega_s} + \frac{Q_p}{\omega_p} \geq 0$ $\frac{Q_p}{\omega_p} + \frac{Q_-}{\omega_-} \geq 0$
Type IV Capacitive	$\frac{Q_s[1 - \cos\theta_s]}{\omega_s} + \frac{Q_p[1 - \cos\theta_p]}{\omega_p} + \frac{Q_-[1 - \cos(\theta_p - \theta_s)]}{\omega_-} \leq 0$ $\frac{Q_s\chi_s^2}{\omega_s} + \frac{Q_p\chi_p^2}{\omega_p} + \frac{Q_-(\chi_p - \chi_s)^2}{\omega_-} \leq 0$ $\frac{Q_s}{\omega_s} + \frac{Q_-}{\omega_-} \leq 0$ $\frac{Q_s}{\omega_s} + \frac{Q_p}{\omega_p} \leq 0$ $\frac{Q_p}{\omega_p} + \frac{Q_-}{\omega_-} \leq 0$

Table B.6. Frequency-power formulas of the four types, with only four frequencies: ω_s, ω_p, $\omega_- = \omega_p - \omega_s$ and $\omega_+ = \omega_p + \omega_s$. Only a few of the possible formulas of types III and IV are given; they hold for all values of the parameters θ_s, θ_p, χ_s and χ_p.

Type I Manley-Rowe Formulas	$\frac{P_s}{\omega_s} - \frac{P_-}{\omega_-} + \frac{P_+}{\omega_+} = 0$ $\frac{P_p}{\omega_p} + \frac{P_-}{\omega_-} + \frac{P_+}{\omega_+} = 0$
Type II	$Q_s - Q_- + Q_+ = 0$ $Q_p + Q_- + Q_+ = 0$
Type III	$P_s[1 - \cos\theta_s] + P_p[1 - \cos\theta_p] + P_-[1 - \cos(\theta_p - \theta_s)] + P_+[1 - \cos(\theta_p + \theta_s)] \geq 0$ $P_s\chi_s^2 + P_p\chi_p^2 + P_-(\chi_p - \chi_s)^2 + P_+(\chi_p + \chi_s)^2 \geq 0$ $P_s + P_- + P_+ \geq 0$ $P_p + P_- + P_+ \geq 0$ $P_s + P_p + 4P_- \geq 0$ $P_s + P_p + 4P_+ \geq 0$

Table B.6.(Cont.)

Type IV Inductive	$\frac{Q_s[1-\cos\theta_s]}{\omega_s}+\frac{Q_p[1-\cos\theta_p]}{\omega_p}+\frac{Q_-[1-\cos(\theta_p-\theta_s)]}{\omega_-}+\frac{Q_+[1-\cos(\theta_p+\theta_s)]}{\omega_+}\geq 0$ $\frac{Q_s\chi_s^2}{\omega_s}+\frac{Q_p\chi_p^2}{\omega_p}+\frac{Q_-(\chi_p-\chi_s)^2}{\omega_-}+\frac{Q_+(\chi_p+\chi_s)^2}{\omega_+}\geq 0$ $\frac{Q_s}{\omega_s}+\frac{Q_-}{\omega_-}+\frac{Q_+}{\omega_+}\geq 0$ $\frac{Q_p}{\omega_p}+\frac{Q_-}{\omega_-}+\frac{Q_+}{\omega_+}\geq 0$ $\frac{Q_s}{\omega_s}+\frac{Q_p}{\omega_p}+4\frac{Q_-}{\omega_-}\geq 0$ $\frac{Q_s}{\omega_s}+\frac{Q_p}{\omega_p}+4\frac{Q_+}{\omega_+}\geq 0$
Type IV Capacitive	$\frac{Q_s[1-\cos\theta_s]}{\omega_s}+\frac{Q_p[1-\cos\theta_p]}{\omega_p}+\frac{Q_-[1-\cos(\theta_p-\theta_s)]}{\omega_-}+\frac{Q_+[1-\cos(\theta_p+\theta_s)]}{\omega_+}\leq 0$ $\frac{Q_s\chi_s^2}{\omega_s}+\frac{Q_p\chi_p^2}{\omega_p}+\frac{Q_-(\chi_p-\chi_s)^2}{\omega_-}+\frac{Q_+(\chi_p+\chi_s)^2}{\omega_+}\leq 0$ $\frac{Q_s}{\omega_s}+\frac{Q_-}{\omega_-}+\frac{Q_+}{\omega_+}\leq 0$ $\frac{Q_p}{\omega_p}+\frac{Q_-}{\omega_-}+\frac{Q_+}{\omega_+}\leq 0$ $\frac{Q_s}{\omega_s}+\frac{Q_p}{\omega_p}+4\frac{Q_-}{\omega_-}\leq 0$ $\frac{Q_s}{\omega_s}+\frac{Q_p}{\omega_p}+4\frac{Q_+}{\omega_+}\leq 0$

Table B.7. Some devices that are known to obey frequency-power formulas of each type. The reference given is to an earlier section of this book. The exact form of the expressions for power at each frequency should be consulted before the formulas are used, especially for distributed devices

Type I (Manley-Rowe Formulas)

Section	Device
6.2	Acoustic system
4.2	Capacitor, linear
4.2	Capacitor, nonlinear
4.4	Capacitor, time-varying
	Circulator
4.3	Coils, coupled
6.1	Dielectric, nonlinear
4.9	Differential gear
6.1	Electromagnetic medium
6.3	Electron beam, irrotational
6.6	Electron beam, rotational
4.7	Energy conversion device, conservative
6.4	Ferrite
6.2	Fluid flow, irrotational
6.6	Fluid flow, rotational
4.9	Gear
4.9	Gyrator
6.4	Gyromagnetic medium
6.2	Hydrodynamics, irrotational
6.6	Hydrodynamics, rotational
4.9	Ideal transformer
4.6	Induction motor
4.3	Inductor, linear
4.3	Inductor, nonlinear
4.4	Inductor, time-varying
4.7	Loudspeaker
6.7	Magnetohydrodynamic system
4.9	Maser model
4.5	Mass
4.5	Mechanical spring
4.7	Microphone
	Moment of Inertia
4.10	Network
4.7	Phonograph cartridge
6.6	Plasma model
4.6	Rotating machine model
	Rotational spring
4.7	Solenoid
4.5	Spring, mechanical
4.8	Traditor
4.9	Transformer, ideal
4.9	Transformer, physical
	Transmission line
	Waveguide

Type II

Section	Device
5.1	Avalanche diode
5.2	Carbon microphone
5.2	Cryotron
5.1	Crystal diode
5.1	Dashpot
5.1	Diode
5.1	Diode, ideal
5.1	Friction
5.1	Ideal diode
5.3	Ideal transformer
5.4	Network
5.2	Piezoresistive device
5.2	Photoconductor
5.1	p-n-p-n diode
5.2	Potentiometer
5.1	Ratchet
5.1	Rectifier
5.1	Resistor, linear
5.1	Resistor, nonlinear
5.2	Resistor, time-varying
5.2	Rheostat
5.1	Semiconductor diode
5.2	Strain gage
5.2	Superconductor
5.2	Switch
5.2	Thermistor
5.3	Transformer, ideal
5.3	Transformer, time-varying
5.1	Tunnel diode
5.1	Varistor

Table B.7 (continued)

Type III		Type IV, Inductive	
5.1	Avalanche diode	4.3	Coils, coupled
5.5	Capacitor, linear	4.9	Ideal transformer
5.2	Carbon microphone	4.3	Inductor, linear
	Circulator	4.3	Inductor, nonlinear
5.2	Cryotron	4.4	Inductor, time-varying
5.1	Crystal diode	4.10	Network
5.1	Dashpot	4.11	Resistor, linear
5.1	Diode	4.7	Solenoid
6.1	Electromagnetic medium, lossy	4.9	Transformer, physical
5.1	Friction		
5.3	Gyrator		
5.1	Hall effect device		
5.1	Ideal diode		
5.3	Ideal transformer		
5.5	Inductor, linear		
5.4	Network		
5.2	Piezoresistive device		
5.2	Photoconductor		
5.2	Potentiometer		
5.1	Ratchet	**Type IV, Capacitive**	
5.1	Rectifier		
5.1	Resistor, linear	4.2	Capacitor, linear
5.1	Resistor, nonlinear	4.2	Capacitor, nonlinear
5.2	Resistor, time-varying	4.4	Capacitor, time-varying
5.2	Rheostat	4.10	Network
5.1	Semiconductor diode	4.11	Resistor, linear
5.2	Strain gage	4.9	Transformer, ideal
5.2	Superconductor		
5.2	Switch		
5.2	Thermistor		
5.3	Transformer, ideal		
5.3	Transformer, time-varying		
	Transmission line		
5.1	Varistor		
	Waveguide		

Table B.8. Some of the more important systems that are known not to obey frequency-power formulas of each type

Type I (Manley-Rowe Formulas)	Type II
Diode Friction Rectifier Resistor Switch Transformer, time-varying	Capacitor Circulator Distributed systems Gyrator Inductor Traditor Transformer, physical Transmission line Waveguide
Type III	**Type IV**
Capacitor, nonlinear Capacitor, time-varying Distributed systems, nonlinear (with exceptions) Inductor, nonlinear Inductor, time-varying Traditor Transformer, physical	Circulator Diode Distributed systems Gyrator Resistor, nonlinear Resistor, time-varying Switch Traditor Transformer, time-varying Transmission line Waveguide

APPENDIX C

We wish to show here that for the purposes of using the frequency-power formulas of the first type, the Manley-Rowe formulas, it is permissible to lump together power at some frequency ω and all its harmonics, and to consider it all as power at the fundamental frequency ω.

If we denote by P_n the power at frequency $n\omega$, the contribution to Eq. 3.36 from frequency ω and all its harmonics is

$$\Sigma_n \frac{P_n}{n\omega} \frac{\partial(n\omega)}{\partial\omega_a} = \Sigma_n \frac{P_n}{n\omega} \frac{n\,\partial\omega}{\partial\omega_a} = \frac{\Sigma_n P_n}{\omega} \frac{\partial\omega}{\partial\omega_a} \tag{C.1}$$

which is precisely the contribution we would get by considering the power $\Sigma_n P_n$ to be put into the system at the fundamental frequency ω. Thus one need not, for the purpose of applying these formulas, distinguish between power at a frequency and power at its harmonics.

Note that this is not true for frequency-power formulas of the other three types.

APPENDIX D

Here we demonstrate a fact of some importance in applying frequency-power formulas of the first type to rotating machines. Generally there is d-c power flow into such machines via the shaft, and a rigorous application of the Manley-Rowe formulas requires the use of the d-c terms. However, a rotating shaft with an average speed ω ought, we feel, to be putting power in at frequency ω, since that is the rate at which the physical elements pass each other. We show here that it is permissible to consider the d-c power input at a shaft as power at frequency ω, provided the shaft angle ϕ is one of the x_i variables that describe the energy state function (as it generally is).

Expressing the shaft angle ϕ in the form of Eq. 3.17, we must equate the v_{i0} quantity with the average shaft speed ω, whether or not the shaft runs at constant speed. Thus ω is some function of the independent frequencies ω_a, and in the rigorous application of the Manley-Rowe formulas the d-c power input term in Eq. 3.36 would be

$$\frac{P}{\omega} \frac{\partial \omega}{\partial \omega_a} \tag{D. 1}$$

where P is the average torque multiplied by ω. However, if we were to consider this power input as occuring at frequency ω, the contribution to Eq. 3.36 from P would be

$$\frac{P}{\omega} \frac{\partial \omega}{\partial \omega_a} \tag{D. 2}$$

which, being the same as Eq. D.1, is correct. We therefore conclude that no error is made, in applying the formulas, if the actual d-c power input at the shaft is considered as input at frequency ω.

Note that this problem does not arise for the other three types of frequency-power formulas.

APPENDIX E

There are many possible sets of independent frequencies one can use in writing frequency-power formulas. As indicated in Sec. 3.1 (The Frequencies), it is desirable to choose the independent frequencies so that as few as possible are necessary to specify the explicit time variation of the system. Even so, there are many possible choices of the remaining ω_a independent frequencies; frequency-power formulas hold for any such set of independent frequencies. In this appendix we determine when new information is obtained by selecting a different set.

Suppose we have two (equally large) sets of independent frequencies on which the explicit time dependence of the system does not depend; let one set be indexed by a, and the other by b. Then each ω_α can be expressed in terms of either set, and

$$\frac{\partial\omega_\alpha}{\partial\omega_b} = \Sigma_a \frac{\partial\omega_\alpha}{\partial\omega_a}\frac{\partial\omega_a}{\partial\omega_b} \tag{E.1}$$

Also, for devices that obey the Manley-Rowe formulas,

$$\frac{\partial v_{i0}}{\partial\omega_b} = \Sigma_a \frac{\partial v_{i0}}{\partial\omega_a}\frac{\partial\omega_a}{\partial\omega_b} \tag{E.2}$$

Frequency-power formulas of Type I, using the ω_a independent frequencies, are

$$\Sigma_i \frac{P_{i0}}{v_{i0}}\frac{\partial v_{i0}}{\partial\omega_a} + \Sigma_\alpha \frac{P_\alpha}{\omega_\alpha}\frac{\partial\omega_\alpha}{\partial\omega_a} \tag{E.3}$$

If each is multiplied by $\partial\omega_a/\partial\omega_b$, the sum of the formulas becomes

$$\Sigma_i \frac{P_{i0}}{v_{i0}}\frac{\partial v_{i0}}{\partial\omega_b} + \Sigma_\alpha \frac{P_\alpha}{\omega_\alpha}\frac{\partial\omega_\alpha}{\partial\omega_b} = 0 \tag{E.4}$$

or simply the Manley-Rowe formula using ω_b as an independent frequency. Thus we can get Manley-Rowe formulas using one set of frequencies from those using any other set, so no new information is obtained by using new independent frequencies.

The same is true of frequency-power formulas of Type II — the same information is contained in the formulas using each set of independent frequencies ω_a.

However, in using frequency-power formulas of Types III and IV, one can sometimes get new information from a new choice of independent frequencies. This occurs when the formulas used are Eqs. 3.52 or 3.53, the most common forms for Type III, and Eqs. 3.67 or 3.68 for Type IV. However, when one uses the more general forms, Eqs. 3.55, 3.56, or 3.58, or Eqs. 3.69, 3.70, or 3.71, then no new information is obtained. To conclude this, for example for Eq. 3.55, merely put the constants θ_a equal to

$$\theta_a = \Sigma_b \frac{\partial\omega_a}{\partial\omega_b} \theta_b \tag{E.5}$$

for any set of constants θ_b. Then Eq. 3.55 becomes

$$\Sigma_a P_a \left[1 - \cos\left(\Sigma_b \frac{\partial\omega_a}{\partial\omega_b} \theta_b \right) \right] \geq 0 \tag{E.6}$$

which is merely the corresponding formula using the ω_b set of independent frequencies.

In summary, new information is obtained by a new choice of independent frequencies ω_a only when using formulas of Types III or IV, and even then only when using the simple forms of Eqs. 3.52, 3.53, 3.67, or 3.68.

APPENDIX F

We examine here some conditions under which

$$\nabla \times \overline{p} = 0 \tag{F.1}$$

for systems with fluid flow, where $\overline{p}$ is the momentum per particle. We show from the equations of motion that if Eq. F.1 is ever satisfied, then under certain conditions it holds for all time. Flow with this property is called "irrotational;" the equations of motion are simplified and, as shown in Secs. 6.2 and 6.3, the Manley-Rowe formulas are more easily obtained.

For simplicity we discuss at the same time both the acoustic system and the electron beam system. To specialize to either case, merely omit the irrelevant terms. To be quite general, we add viscous force to the general force equation of Sec. 6.7, Eq. 6.210:

$$\nabla\left(T + \Upsilon + \frac{\pi}{\rho} + \Omega\right) = -\dot{\overline{p}} + q\dot{\overline{A}} + q\overline{E} + \overline{v} \times \overline{u} + \theta \nabla s + \frac{\rho_f \overline{E}}{\rho} + \frac{\overline{J}_f \times \overline{B}}{\rho} + \overline{F}_v \tag{F.2}$$

where for simplicity we define

$$\overline{u} = \nabla \times \overline{p} \tag{F.3}$$

The force equation for fluid flow, Eq. 6.42, is obtained by dropping the terms involving Ω, $\overline{A}$, $\overline{E}$, $\overline{B}$, θ, s, ρ_f, and $\overline{J}_f$ from Eq. F.2; the electron beam force equation, Eq. 6.74, is obtained by omitting the terms involving Υ, π, Ω, θ, s, ρ_f, $\overline{J}_f$, and $\overline{F}_v$.

We take the curl of each side of Eq. F.2, obtaining

$$\dot{\overline{u}} = \nabla \times (\overline{v} \times \overline{u}) + \nabla \times \overline{F}_v + \nabla\theta \times \nabla s + \nabla \times \frac{\rho_f \overline{E} + \overline{J}_f \times \overline{B}}{\rho} \tag{F.4}$$

or, using a vector identity for $\nabla \times (\overline{v} \times \overline{u})$,

$$\frac{D\overline{u}}{Dt} = \dot{\overline{u}} + (\overline{v} \cdot \nabla)\overline{u}$$

$$= (\overline{u} \cdot \nabla)\overline{v} - \overline{u}(\nabla \cdot \overline{v}) + \nabla \times \overline{F}_v + \nabla\theta \times \nabla s + \nabla \times \frac{\rho_f \overline{E} + \overline{J}_f \times \overline{B}}{\rho} \tag{F.5}$$

If there is no viscous force, and if either the temperature or the entropy is uniform, and if there are no free charges or currents, then Eq. F.5 reduces to

$$\frac{D\overline{u}}{Dt} = (\overline{u} \cdot \nabla)\overline{v} - \overline{u}(\nabla \cdot \overline{v}) \tag{F.6}$$

For any given particle, this is a set of three coupled linear total differential equations, with time-varying coefficients. For given initial conditions, there is a unique solution. But by inspection a possible solution is the trivial one $\overline{u} = 0$; therefore if a particle has zero curl of its momentum at any time, it always does. It makes sense, then, to talk about "irrotational flow," because if the flow is ever irrotational, it remains so.

On the other hand, if there are viscous forces, or if the entropy and the temperature are both nonuniform, or if there are free charges and currents, then Eq. F.6 does not hold. These effects spontaneously introduce a curl of the momentum; if any of them is present, it does not make sense to speak of irrotational flow.

Eq. F.6 is the differential form of Lagrange's Theorem [28; 101, Sec. 2.3], which is usually written in terms of a surface integral of $\bar{u}$ being an invariant of the motion.

Nonviscous fluid flow is irrotational if it starts from irrotational motion, for example if it starts from rest. An electron beam is irrotational if the electrons come from a source that is constructed properly. Gabor [28] has shown that electrons coming from a cathode have only a normal component of $\bar{u}$, that is, a component perpendicular to the surface of the cathode, when they leave. Furthermore, even this component vanishes if there is no magnetic field normal to the cathode. Thus, at any point in space occupied by electrons that have come from a cathode with no normal magnetic field, $\bar{u} = 0$. Any electron beam device with a magnetically-shielded cathode therefore has irrotational flow [28].

REFERENCES

1. Adler, R., G. Hrbek, and G. Wade, "A Low-Noise Electron-Beam Parametric Amplifier," Proc. IRE, 46, 10, 1756-1757 (1958).
2. Bloembergen, N., "Magnetic Resonance in Ferrites," Proc. IRE, 44, 10, 1259-1269 (1956).
3. Bobroff, D. L., "Independent Space Variables for Small-Signal Electron Beam Analyses," IRE Trans. on Electron Devices, ED-6, 1, 68-78 (1959).
4. Bobroff, D. L. and H. A. Haus, "Uniqueness and Orthogonality of Small Signal Solutions in Electron Beams," Res. Div., Raytheon Manufacturing Co., Waltham, Mass., Tech. Rep. No. 31 (1958).
5. Boot, H. A. H., S. A. Self, and R. B. R-Shersby-Harvie, "Containment of a Fully-Ionized Plasma by Radio-Frequency Fields," J. Elect. and Control, 4, 5, 434-453 (1958).
6. Bridges, T. J., "A Parametric Electron Beam Amplifier," Proc. IRE, 46, 2, 494-495 (1958).
7. Brillouin, L., Wave Propagation in Periodic Structures, Dover Publications, New York (1953).
8. Buck, D. A., "The Cryotron — A Superconductive Computer Component," Proc. IRE, 44, 4, 482-493 (1956).
9. Burns, F. P., "Piezoresistive Semiconductor Microphone," J. Acoust. Soc., 29, 2, 248-253 (1957).
10. Callen, H. B., "A Note on the Manley-Rowe Energy Relations for Parametric Amplifiers," J. Franklin Inst., 269, 2, 93-96 (1960).
11. Carlin, H. J., "The Scattering Matrix in Network Theory," IRE Trans. on Circuit Theory, CT-3, 2, 88-97 (1956).
12. Carroll, J. E., "A Simplified Derivation of the Manley and Rowe Power Relationships," J. Elect. and Control, 6, 4, 359-361 (1959).
13. Chang, K. K. N., "Harmonic Generation with Nonlinear Reactances," RCA Review, 19, 3, 455-464 (1958).
14. Cherry, C., "Some General Theorems for Non-Linear Systems Possessing Reactance," Phil. Mag., 42, 333, 1161-1177 (1951).
15. Chu, L. J., "A Kinetic Power Theorem," 1951 IRE Conference on Electron Devices, Durham, N.H., June, 1951.
16. Clavier, P. A., "The Manley-Rowe Relations," Proc. IRE, 47, 10, 1781-1782 (1959).
17. Cowling, T. G., Magnetohydrodynamics, Interscience Publishers, New York (1957).
18. Duinker, S., "General Energy Relations for Parametric Amplifying Devices," Tijdschrift van het Nederlands Radiogenootschap, 24, 5, 287-310 (1959).

19. Duinker, S., "General Properties of Frequency-Converting Networks," Philips Research Reports, 13, 1, 37-78 (1958) and 13, 2, 101-148 (1958).

20. Duinker, S., "Traditors, a New Class of Non-Energic Non-Linear Network Elements," Philips Research Reports, 14, 1, 29-51 (1959).

21. Erdelyi, E., E. E. Kolatorowicz, and W. R. Miller, "The Limitations of Induction Generators in Constant-Frequency Aircraft Systems," Trans. AIEE, 77, Part II (Applications and Industry), 348-352 (1958).

22. Esaki, L., "New Phenomenon in Narrow Germanium p-n Junctions," Phys. Rev., 109, 2, 603-604 (1958).

23. Fano, R. M., L. J. Chu, and R. B. Adler, Electromagnetic Fields, Energy, and Forces, John Wiley & Sons, New York (1960).

24. Faraday, M., "On a Peculiar Class of Acoustical Figures; and on Certain Forms Assumed by Groups of Particles upon Vibrating Elastic Surfaces," Phil. Trans. of the Roy. Soc. (London), 121, Part II, 299-340 (1831); and appendix, "On the Forms and States Assumed by Fluids in Contact with Vibrating Elastic Surfaces," 319.

25. Fitzgerald, A. E. and C. Kingsley, Jr., Electric Machinery, McGraw-Hill Book Co., New York (1952).

26. Fontana, J. R. and H. J. Shaw, "Harmonic Generation at Microwave Frequencies Using Field-Emission Cathodes," Proc. IRE, 46, 7, 1424-1425 (1958).

27. Franks, L. E. and F. J. Witt, "Solid-State Sampled-Data Bandpass Filters," International Solid-State Circuits Conference, Philadelphia, Feb. 10-12, 1960.

28. Gabor, D., "Dynamics of Electron Beams," Proc. IRE, 33, 11, 792-805 (1945).

29. Gardner, M. F. and J. L. Barnes, Transients in Linear Systems, John Wiley & Sons, New York, 69-70 (1942).

30. Goldstein, H., Classical Mechanics, Addison-Wesley Publishing Co., Reading (Mass.) (1950).

31. Goto, E., "The Parametron, a Digital Computing Element Which Utilizes Parametric Oscillation," Proc. IRE, 47, 8, 1304-1316 (1959).

32. Grubbs, W. J., "The Hall Effect Circulator – A Passive Transmission Device," Proc. IRE, 47, 4, 528-535 (1959).

33. Guillemin, E. A., Introductory Circuit Theory, John Wiley & Sons, New York (1953).

34. Guillemin, E. A., "Some Generalizations of Linear Network Analysis That Are Useful when Active and/or Nonbilateral Elements are Involved," M. I. T. Research Laboratory of Electronics Quarterly Progress Report, 103-114 (Oct. 15, 1957).

35. Haus, H. A., "Power-Flow Relations in Lossless Nonlinear Media," IRE Trans. on Microwave Theory and Techniques, MTT-6, 3, 317-324 (195

36. Haus, H. A., "The Kinetic Power Theorem for Parametric, Longitudinal, Electron-Beam Amplifiers," IRE Trans. on Electron Devices, ED-5, 4, 225-232 (1958).

37. Haus, H. A., "On a Nonlinear Phenomenon in Plasmas," M. I. T. Electrical Engineering Department, Energy Conversion Group Unpublished Internal Memorandum No. 2 (Mar. 27, 1959).

38. Haus, H. A., private communication (May, 1959).

39. Haus, H. A. and D. L. Bobroff, "Small Signal Power Theorem for Electron Beams," J. Appl. Phys., 28, 6, 694-704 (1957).

40. Haus, H. A. and P. Penfield, Jr., "On the Noise Performance of Parametric Amplifiers," M. I. T. Electrical Engineering Department, Energy Conversion Group Unpublished Internal Memorandum No. 19 (Aug. 11, 1959).

41. Haus, H. A. and F. N. H. Robinson, "The Minimum Noise Figure of Microwave Beam Amplifiers," Proc. IRE, 43, 8, 981-991 (1955).

42. Hildebrand, F. B., Advanced Calculus for Engineers, Prentice-Hall, New York (1949).

43. Hildebrand, F. B., Methods of Applied Mathematics, Prentice-Hall, Englewood Cliffs (1952).

44. Howard, L. N., "Cross Waves," Bull. Am. Phys. Soc., II, 5, 2, 132, (1960).

45. Itô, H., "Variational Principle in Hydrodynamics," Prog. Theo. Phys., 9, 2, 117-131 (1953).

46. Kaufman, I. and D. Douthett, "Harmonic Generation Using Idling Circuits," Proc. IRE, 48, 4, 790-791 (1960).

47. Kino, G. S. and B. Ludovici, "A Plasma Parametric Amplifier," Bull. Am. Phys. Soc., II, 5, 4, 318-319 (1960).

48. Klüver, J. W., "Small Signal Power Conservation Theorem for Irrotational Electron Beams," J. Appl. Phys., 29, 4, 618-622 (1958).

49. Kurokawa, K. and J. Hamasaki, "An Extension of the Mode Theory to Periodically Distributed Parametric Amplifiers with Losses," IRE Trans. on Microwave Theory and Techniques, MTT-8, 1, 10-18 (1960).

50. Kurokawa, K. and J. Hamasaki, "Mode Theory of Lossless Periodically Distributed Parametric Amplifiers," IRE Trans. on Microwave Theory and Techniques, MTT-7, 3, 360-365 (1959).

51. Lamb, H., Hydrodynamics, Dover Publications, New York (1945).

52. Leenov, D. and A. Uhlir, Jr., "Generation of Harmonics and Subharmonics at Microwave Frequencies with P-N Junction Diodes," Proc. IRE, 47, 10, 1724-1729 (1959).

53. Leeson, D. B. and S. Weinreb, "Frequency Multiplication with Nonlinear Capacitors — A Circuit Analysis," Proc. IRE, 47, 12, 2076-2084 (1959).

54. Lin, C.-C., private communication (1959).

55. Louisell, W. H., "A Three-Frequency Electron Beam Parametric Amplifier and Frequency Converter," J. Elect. and Control, 6, 1, 1-25 (1959).

56. Louisell, W. H. and C. F. Quate, "Parametric Amplification of Space Charge Waves," Proc. IRE, 46, 4, 707-716 (1958).

57. Low, F. E., "A Lagrangian Formulation of the Boltzmann-Vlasov Equation for Plasmas," Proc. Roy. Soc., A 248, 1253, 282-287 (1958).

58. Manley, J. M., "Some General Properties of Magnetic Amplifiers," Proc. IRE, 39, 3, 242-251 (1951).

59. Manley, J. M. and H. E. Rowe, "General Energy Relations in Nonlinear Reactances," Proc. IRE, 47, 12, 2115-2116 (1959).

60. Manley, J. M. and H. E. Rowe, "Some General Properties of Nonlinear Elements. I. General Energy Relations," Proc. IRE, 44, 7, 904-913 (1956).

61. Mason, W. P., W. H. Hewitt, and R. F. Wick, "Hall Effect Modulators and 'Gyrators' Employing Magnetic Field Independent Orientations in Germanium," J. Appl. Phys., 24, 2, 166-175 (1953).

62. Mawardi, O. K., "On the Concept of Coenergy," J. Franklin Inst., 264, 4, 313-332 (1957).

63. McAfee, K. B., E. J. Ryder, W. Shockley, and M. Sparks, "Observations of Zener Current in Germanium p-n Junctions," Phys. Rev., 83, 3, 650-651 (1951).

64. McWhorter, A. L., private communication (January 11, 1960).

65. Melde, F., Annalen der Physik und Chemie, Poggendorff, ed. (Pogg. Ann.), 109, 193-215 (1860) and 111, 513-537 (1860).

66. Messenger, G. C. and C. T. McCoy, "Theory and Operation of Crystal Diodes as Mixers," Proc. IRE, 45, 9, 1269-1283 (1957).

67. Mikoshiba, N., "Quantum Hydrodynamics for a Charged Non-Viscous Fluid," Prog. Theo. Phys., 13, 6, 627-628 (1955).

68. Millar, W., "Some General Theorems for Non-Linear Systems Possessing Resistance," Phil. Mag., 42, 333, 1150-1160 (1951).

69. Moll, J. L., M. Tanenbaum, J. M. Goldey, and N. Holonyak, "P-N-P-N Transistor Switches," Proc. IRE, 44, 9, 1174-1182 (1956).

70. Morse, P. M. and H. Feshbach, Methods of Theoretical Physics, McGraw-Hill Book Co., New York (1953).

71. Mumford, W. W., "Some Notes on the History of Parametric Transducers," Proc. IRE, 48, 5, 848-853 (1960).

72. Olson, F. A., C. P. Wang, and G. Wade, "Parametric Devices Tested for Phase-Distortionless Limiting," Proc. IRE, 47, 4, 587-588 (1959).

73. Page, C. H., "Frequency Conversion with Nonlinear Reactance," J. Research of the National Bureau of Standards, 58, 5, 227-236 (1957).

74. Page, C. H., "Frequency Conversion with Positive Nonlinear Resistors," J. Research of the National Bureau of Standards, 56, 4, 179-182 (1956).

75. Page, C. H., "Harmonic Generation with Ideal Rectifiers," Proc. IRE, 46, 10, 1738-1740 (1958).

76. Panofsky, W. K. H. and M. Phillips, Classical Electricity and Magnetism, Addison-Wesley Publishing Co., Reading (Mass.) (1955).

77. Pantell, R. H., "General Power Relationships for Positive and Negative Nonlinear Resistive Elements," Proc. IRE, 46, 12, 1910-1913 (1958).

78. Penfield, P., Jr., "Generalization of the Frequency-Power Formulas of Manley and Rowe," Proc. Polytechnic Institute of Brooklyn Symposia, 10, Polytechnic Institute of Brooklyn, Brooklyn, N.Y. (1960).

79. Penfield, P., Jr., A. V. Oppenheim, and R. F. Webber, "A Rotating Machine Parametric Amplifier," M.I.T. Electrical Engineering Department, Energy Conversion Group Unpublished Internal Memorandum No. 1 (Mar. 27, 1959).

80. Penhune, J. P., "A General Condition for the Formation of a State Function," M.I.T. Electrical Engineering Department, Energy Conversion Group Unpublished Internal Memorandum No. 5 (Apr. 1, 1959).

81. Pittman, P. F., "The Application of the Dynistor Diode to 'Off-On' Controllers," Semiconductor Products, 2, 3, 23-26 (1959).

82. Rayleigh, Lord, "On Maintained Vibrations," Phil. Mag. 15, 94, 229-235 (1883).

83. Rayleigh, Lord, "On the Maintenance of Vibration by Forces of Double Frequency and on the Propagation of Waves through a Medium Endowed with a Periodic Structure," Phil. Mag., 24, 147, 145-159 (1887).

84. Rediker, R. H. and D. E. Sawyer, "Very Narrow Base Diode," Proc. IRE, 45, 7, 944-953 (1957).

85. Riaz, M., "Characteristics of Induction Machines in Variable-Speed Constant-Frequency Generating Systems," W.A.D.C. Technical Note TN 58-104, Wright Air Development Center, Wright-Patterson Air Force Base, Ohio (Apr. 1958).

86. Riaz, M., "Energy Conversion Properties of Induction Machines in Variable-Speed Constant-Frequency Generating Systems," AIEE paper no. CP 58-917, delivered at the AIEE Summer General Meeting and Air Transportation Conference, Buffalo, N.Y., June 22-27, 1958.

87. Robbins, L., "The A-C Potentiometer — A New Circuit Component," Electronic Industries, 19, 3, 120-124 (1960).

88. Rostoker, N., "Plasma Stability with ac Confining Fields," Bull. Am. Phys. Soc., II, 5, 4, 307 (1960).

89. Rowe, H. E., "Some General Properties of Nonlinear Elements. II. Small Signal Theory," Proc. IRE, 46, 5, 850-860 (1958).

90. Salzberg, B., "Masers and Reactance Amplifiers — Basic Power Relations," Proc. IRE, 45, 11, 1544-1545 (1957).

91. Serrin, J., private letter to C.-C. Lin (Aug. 23, 1958). This letter was shown to me though the courtesy of Professor Lin.

92. Serrin, J., "Mathematical Principles of Classical Fluid Mechanics," Handbuch der Physik, VIII/1, 125-263, Springer-Verlag, Berlin (1959).

93. Shafer, C. G., "Noise Figure for a Traveling-Wave Parametric Amplifier of the Coupled-Mode Type," Proc. IRE, 47, 12, 2117-2118 (1959).

94. Shekel, J., "The Gyrator as a 3-Terminal Element," Proc. IRE, 41, 8, 1014-1016 (1953).

95. Shockley, W., "Unique Properties of the Four-Layer Diode," Electronic Industries, 16, 8, 58-60, 161, 162, 163, 164, 165 (1957).

96. Shockley, W. and J. F. Gibbons, "Introduction to the Four-Layer Diode," Semiconductor Products, 1, 1, 9-13 (1958).

97. Sommers, H. S., Jr., "Tunnel Diodes as High-Frequency Devices," Proc. IRE, 47, 7, 1201-1206 (1959).

98. Spencer, E. G. and R. C. LeCraw, "Magnetoacoustic Resonance in Yttrium Iron Garnet," Phys. Rev. Letters, 1, 7, 241-243 (1958).

99. Sturrock, P.A., "Action-Transfer and Frequency-Shift Relations in the Nonlinear Theory of Waves and Oscillations," Ann. Physics, 9, 3, 422-434 (1960).

100. Sturrock, P. A., "A Variational Principle and an Energy Theorem for Small-Amplitude Disturbances of Electron Beams and of Electron-Ion Plasmas," Ann. Physics, 4, 3, 306-324 (1958).

101. Sturrock, P.A., Static and Dynamic Electron Optics, Cambridge University Press, Cambridge (England) (1955).

102. Suhl, H., "The Nonlinear Behavior of Ferrites at High Microwave Signal Levels," Proc. IRE, 44, 10, 1270-1284 (1956).

103. Suhl, H., "Theory of the Ferromagnetic Microwave Amplifier," J. Appl. Phys., 28, 11, 1225-1236 (1957).

104. Tellegen, B. D. H., "The Gyrator, A New Electric Network Element," Philips Research Reports, 3, 2, 81-101 (1948).

105. Thellung, A., "On the Hydrodynamics of Non-Viscous Fluids and the Theory of Helium II. Part II," Physica, 19, 217-226 (1953).

106. Torrey, H. C. and C. A. Whitmer, Crystal Rectifiers, McGraw-Hill Book Co., New York; M.I.T. Radiation Laboratory Series, 15 (1948).

107. Uhlir, A., Jr., "The Potential of Semiconductor Diodes in High-Frequenc Communications," Proc. IRE, 46, 6, 1099-1115 (1958).

108. Uhlir, A., Jr., "Two-Terminal P-N Junction Devices for Frequency Conversion and Computation," Proc. IRE, 44, 9, 1183-1191 (1956).

109. von Neumann, J., "Non-linear Capacitance or Inductance Switching, Amplifying, and Memory Organs," U.S. Patent No. 2,815,488, issued Dec. 3, 1957, assigned to the I.B.M. Corporation.

110. Wagner, R. R., "A Lossless Acoustic Power Theorem," B.S. Thesis, Department of Electrical Engineering, M.I.T. (May, 1959).

111. Warner, R. M., Jr., W. H. Jackson, E. I. Doucette, and H. A. Stone, Jr., "A Semiconductor Current Limiter," Proc. IRE, 47, 1, 44-56 (1959).

112. Wei, C.-C., "Relativistic Hydrodynamics for a Charged Nonviscous Fluid," Phys. Rev., 113, 6, 1414 (1959).

113. Weibel, E. S., "Confinement of a Plasma Column by Radiation Pressure, The Plasma in a Magnetic Field, ed. by R. K. M. Landshoff, Stanford University Press, Stanford (1958).

114. Weiss, M. T., "A Solid-State Microwave Amplifier and Oscillator Using Ferrites," Phys. Rev., 107, 1, 317 (1957).

115. Weiss, M. T., "Quantum Derivation of Energy Relations Analogous to Those for Nonlinear Reactances," Proc. IRE, 45, 7, 1012-1013 (1957).

116. White, D. C. and H. H. Woodson, Electromechanical Energy Conversion, John Wiley & Sons, New York (1959).

117. Wigington, R. L., "A New Concept in Computing," Proc. IRE, 47, 4, 516-523 (1959).

118. Woodford, J. B., Jr. and D. L. Feucht, "The Superconductive Transition Radio-Frequency Mixer and the Problem of Cryotron Switching Time," Proc. IRE, 46, 11, 1871 (1958).

119. Wuerker, R. F., H. Shelton, and R. V. Langmuir, "Electrodynamic Containment of Charged Particles," J. Appl. Phys., 30, 3, 342-349 (1959).

120. Yeh, C., "Generalized Energy Relations of Nonlinear Reactive Elements," Proc. IRE, 48, 2, 253 (1960).

121. Ziman, J. M., "Quantum Hydrodynamics and the Theory of Liquid Helium," Proc. Roy. Soc., A 219, 1137, 257-270 (1953).

INDEX